Elegant and Easy

Elegant and Easy

Decorative Ideas for Food Presentations

Jean F. Nicolas

CBI Publishing Company, Inc.
286 Congress Street
Boston, MA 02210

Dedication

To Mrs. Charles W. Engelhard, to whom I will remain grateful forever, for inspiring me to cook, decorate, and serve food with beauty and simplicity.

Cover design, text design, and production: Martucci Studio Book Division.
Illustrations: Joseph Farnham.

Copyright © 1983 by CBI Publishing Company, Inc. All rights reserved. This book may not be reproduced by any means without written permission from the publisher.

Printed in the United States of America.

Printing (last digit): 9 8 7 6 5 4 3 2 1

Library of Congress Cataloging in Publication Data

Nicolas, Jean F.
    Elegant and easy.

  1. Food Presentation.   2. Cookery (Garnishes).
3. Fruit.   4. Vegetables.   I. Title.
TX652.N47   1983   641.6′4   82-20699
ISBN 0-8436-2266-0
ISBN 0-8436-2267-9 (pbk.)

Contents

1 Introduction 1
Tools Used in Food Decorating   2

2 Fruit Decorations 5
Apples   7
   The Apple Basket   8
   The Apple Bird of Paradise   9
   The Sawtooth Apple   11
   The Apple Tulip   12
   Two Apples in One   13
   V-Shaped Apple Slices   15
Avocados   16
Citrus Fruits   17
   Grooved Lemon Slices   18
   The Lemon Basket   19
   Lemon and Orange Wedges   20
   The Lemon Tulip Cup   21
Grapes   22
   Sugar-Frosted Grapes   23
Melon   24
   The Honeydew Basket   25
   The Melon Cradle   27
   The Honeydew Melon Swan   28
   Papaya and Blueberries   30
Pears   31
   Pear Mice   31
Pineapple   32
   The Pineapple Bird   32
   The Pineapple Cart   33

3 Vegetable Decorations 35

Carrots 36
- *Carrot Balls* 37
- *The Carrot Daisy* 38
- *The Carrot Flower Cup* 39
- *The Carrot Lily* 41
- *Carrot Pinwheels* 43
- *Carrot Stick Bundles* 44
- *Corrugated Carrot Slices* 45
- *The Palm Tree* 46
- *The Spider Mum* 48

Cucumbers 50
- *The Cucumber Basket* 50
- *The Cucumber Chain* 52
- *Cucumber Fish Scales* 53
- *The Cucumber Tulip Cup* 54
- *The Cucumber Viking Ship* 55

More Ideas Using Cucumbers 56
Belgian Endive 57
Leeks 58
- *The Leek Blossom* 58

More Decorations Using Leeks 60
Field Mushrooms 60
- *Fluted Mushrooms* 61

Olives 61
- *The Olive Chain* 62

Parsley 62
Peppers 63
- *Pepper Baskets* 63

Pickles 64
- *Pickle Rings* 64
- *Pickle Fans* 65

Potatoes 66
- *Potato Roses* 66
- *Fried Potato Nests* 68

Radishes 70
- *The Radish Daisy* 70
- *The Radish Fan* 72

The Radish Rose   73
The Radish Tulip   74
The Red Radish Blossom   75
The White Radish Spray   76
Squash   77
Tomatoes   79
Tomato Aspic Wedges   79
Tomato Mushrooms on a Cracker   80
The Tomato Poinsettia   81
The Tomato Rose   82

4 Sheets of Color   85

Color Sheets   85
Using Color Sheets   85
Preparing the Designs   86

5 Fanciful Meringues   89

Preparing Meringues   89
Tools Required for Meringue Decorations   90
Meringue Fancies   90
The Exotic Meringue   91
Mushrooms on a Bed of Moss   91
Meringue Swans   93
Chocolate Meringue Designs   94
Coconut Meringue Nests   95
Meringue Rocks   95
Meringue Baskets   96
Molded Meringues   96
Meringue Sandwiches   97
Meringue Shells   97

6 More Food Decorations   101

Butter and Cream Cheese Creations   101
Cream Cheese Roses   101
Butter Decorations   102
Bread Decorations   102
Bread Crepe Pans   103
Bread Basket and Horn of Plenty   104

      A Cornucopia   104
      Bread Baskets   106
Eggs in Food Decorating   107
      The 100-Year-Old Eggs   107
      Egg Mimosa   108
      Egg Slices   108
      The Chinese Farmer   108
      The Snowman   109
      Stuffed Whole Eggs   109
      The Lady Bug   109
More Decorative Ideas   110
      The Liver Pâté Cone   110
      Snow Peas Stuffed with Cream Cheese   110
      Thanksgiving Turkey Appetizer Tray   112

7   Bouquets and Containers   113
      A Bouquet of Daisies   113

8   The Picture Plate   117

Acknowledgments

For their generous assistance, I would like to express my appreciation to Bridget O'Reilly, Jane Robertson, and Rita Sternick.

Special thanks to Gunnar, of Gunnar Photo Studio, Bernardsville, N.J. for his photographic assistance, and to Joyce La Capra, owner of the Kitchen Door in Mendham, N.J. for contributing to the selection of the tools used in this book.

Finally, I wish to thank my wife Chantal for providing her intuition and valuable knowledge to the world of artistry in the kitchen.

1
Introduction

Cooking is an art that in the United States has gained the widespread respect it has always been accorded abroad. Among topics of conversation, food is always high on the list. We now realize that a balanced diet, with particular emphasis on a variety of unprocessed, natural foods that are low in carbohydrates, is essential for good nutrition.

In the 1970s, connoisseurs of nouvelle cuisine tried to revolutionize the basics of cooking. Today, nouvelle cuisine has lost much of its popularity, but its lasting contribution is that it encouraged many professionals to reevaluate standards of quality and freshness, and to concentrate on the presentation of food.

Many food experts now go a step further, demonstrating creativity and originality in the presentation of food. Well-prepared foods, attractively decorated and presented are part of the standard repertoire of any successful establishment. The most renowned restaurants in the world now advocate simplicity in cooking and decorating gourmet dishes. In his book *The Banquet Business*, Arno Schmidt, former Executive Chef at the Waldorf Astoria in New York City sums up contemporary thinking about menu planning. "... A contrast of texture, color, and flavor.... There should be a pleasant contrast from dish to dish as well as on each of the plates. Visualize how the food items will look together when served. There should also be a contrast in texture and cooking methods as well."

Without going to the extent of designing menus that require a great deal of planning and expertise, more individual and innovative ideas should be used when decorating food, an aspect of food preparation that, regrettably, is too often neglected.

Many cultures, particularly French, have contributed immensely to transforming our basic need for food into a pleasurable experience. French culinary art associated with Escoffier, Carême, Vatel and many others has never really been surpassed. The artistic abilities of many chefs and cooks were amazing. Every preparation was a work of art, a marvel. However, many eccentric decorative methods were used to achieve the dazzling presentations. Calves feet were transformed into gelatin glazes. Beef fat was rendered and shaped to imitate sculpture. Spectacular ice carvings added translucent beauty to many cold desserts. Truffles, called black diamonds, played an important role in French and Italian cuisine. Unfortunately, many dishes served were decorated to such an extent that they lost their quality and palatability due to overhandling and lack of refrigeration.

Today, food decoration is different. For economical reasons and due to the desire for fresh foods, antiquated decorating methods are disregarded for the most part. Quality, good preparation, and final artistic touches that are simple, effective, economical, and speedy are the order of the day.

Cooking and decorating food, though closely related, are different. Decorations made of vegetables and fruits enable us to vary a menu and to present a simple dish with a bit of elegance. Carrots can be cut into flowers in the same amount of time required to cut them into sticks. Fruit can be served in a swan fashioned from a melon. With a simple tool a radish becomes a flower that adds glamour to a salad or a platter of cold meats. A cucumber boat or a grapefruit basket can be used as sauce boats. These decorations, and more, are easy to create. Painting a dish into a picture is nothing new; the methods have been simplified to meet the needs of today.

We all know that preparing food well requires a lot of time, whether it be a simple country dish, a gourmet dinner, or a banquet. In spite of the emphasis on fresh, natural, quality foods, presentation sometimes leaves a lot to be desired. Some time and thought should always be allowed for decoration and presentation. Any dish deserves a crowning touch, and this book will be a valuable asset to anyone interested in cooking and decorating food.

Tools Used in Food Decorating

As in most professions, special tools are essential to create various food decorations. However, most of the everyday kitchen tools are useful for food decorating. Only a few are specially designed to create specific food decorations. The following drawings illustrate the tools used in this book.

1. mandoline
2. large corrugated vegetable slicer
3. small corrugated vegetable slicer
4. French knife (8 in.)
5. paring knife ($4\frac{1}{2}$ in.)
6. paring knife ($3\frac{1}{2}$ in.)
7. paring knife ($2\frac{1}{2}$ in.)
8. standard Parisian scoop
9. tiny Parisian scoop
10. zesteur or cester to remove fine strips of fruits or vegetables (or lemon peeler)
11. olive-shaped scoop
12. zesteur for carving fruit
13. corrugated knife for slicing fruits or vegetables
14. fruit corer

INTRODUCTION 3

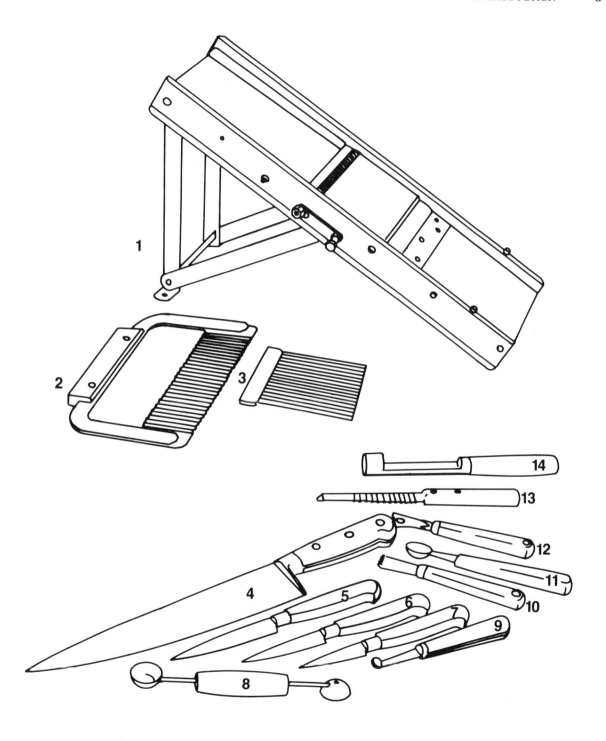

CHAPTER ONE

1. plain round cutter set
2. basket for potato nests
3. set of pastry tubes
4. daisy cutter
5. 12-section wedger for apples, pears, etc.
6. 20-section wedger for radish flowers
7. set of fancy cutters

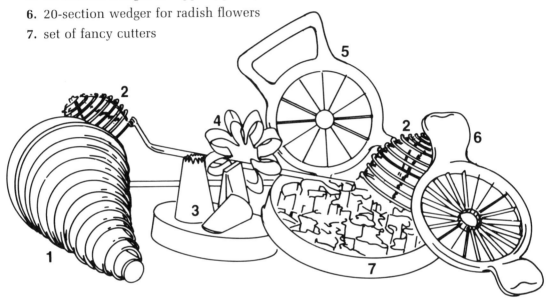

2
Fruit Decorations

Nature is here to charm our eyes and has so much to offer for our health and well-being. We will learn to take advantage of the contrasting colors of fruits and vegetables. The objective of all the decorations presented in this chapter is to embellish and add life to everyday food we prepare, serve, and eat.

Here is a guide to selection and care of fresh fruits used for decorating food.

- **Apples** Select firm apples that have good color for the variety. Any apple can be used for any purpose, but results vary. Keep cold and humid. Available year round.
- **Avocados** Color ranges from purple black to green according to variety. Irregular brown marks on surface are superficial and do not affect the quality. Hold at room temperature until fruit yields gently to pressure, then refrigerate. Available year round.
- **Berries** Choose plump, firm, full colored berries. All varieties, with the exception of strawberries, should be free of hulls. Avoid baskets showing signs of bruised or leaking fruits. Cover and refrigerate. Use within a few days. Available mainly June to August.
- **Cantaloupes** Cream-colored "netting" should completely cover the melon. It should be free of any stem and "give" when pressed gently. Hold at room temperature for a few days, then refrigerate and use as soon as possible. Available May to September.
- **Cherries** Sweet cherries are bright and glossy, ranging from deep red to black in color. They should be attached to fresh green stems. Avoid cherries that are hard, sticky, or light in color. Refrigerate and use within a few days. Available May to August.
- **Grapefruit** Should be firm, not puffy or loose-skinned. Look for globular fruits that are heavy for their size, indicating juiciness. Russeting, cinnamon skin coloration, or green tinge do not affect eating quality. Refrigerate or keep at room temperature. Available year round.
- **Grapes** Choose plump, well colored grapes that are firmly attached to green, pliable stems. Green grapes are sweetest when yellow green in color. Red grapes are best when rich, red color predominates. Grapes will not increase in sweetness, so there is no need to hold them for further ripening. Refrigerate and use within one week. Peak supply is July to November.
- **Honeydew Melons** Look for a creamy or yellowish-white rind with a velvety feel. Avoid stark-white or greenish tinged rinds. Hold at room temperature for a few days, then refrigerate. Peak supply is June to October.

- **Lemons and Limes** Look for fine-textured skin indicating juiciness. Select those that are heavy for their size. Keep at room temperature or refrigerate. Available year round.
- **Oranges** Fruit should be firm and heavy with a fine-textured skin. A green skin color or russeting does not affect eating quality. Store at room temperature or refrigerate. Available year round.
- **Papayas** Select medium size, well colored fruit, that is at least half yellow. Ripen at room temperature until skin color is primarily golden, then refrigerate and use as soon as possible. Peak supply is May to June, October to December.
- **Peaches and Nectarines** Glowing blush is not a true indication of ripeness. Background color of peaches should be yellowish or cream. Nectarines are yellow-orange when ripe. Fruit should be firm with a slight softening along the "seam." Avoid green or green tinged fruits and any that are hard, dull, or bruised. Hold at room temperature to soften, then refrigerate and use promptly. Peak supply is June to September.
- **Pears** Color varies according to variety. Cinnamon russeting on surface does not affect quality. Pears generally require additional ripening. Hold at room temperature until stem end yields to gentle pressure, then refrigerate. Year round availability due to different varieties.
- **Pineapples** Select large fruit that has fresh green leaves. Shell color is not an indicator of maturity. Pineapples do not ripen after harvest, so they may be eaten immediately. Keep at room temperature or refrigerate. Available year round. Peak supply is March to June.
- **Strawberries** Choose berries that are fresh, clean, bright, and red. The green caps should be intact and the fruit should be free of bruises. Strawberries are best eaten immediately. If they must be stored, refrigerate them with their caps intact. Available year round with peak supply April through June.
- **Tangelos** Look for firm, thin-skinned fruits that are heavy for their size. Keep at room temperature or refrigerate. Available October to January.
- **Tangerines** Choose fruit that is heavy for its size. A puffy appearance and feel is normal. Refrigerate and use as soon as possible. Peak supply is November to January.
- **Watermelons** It is difficult to determine ripeness of uncut fruit. Choose firm, smooth melons with a waxy bloom or dullness on the rind. Underside should be yellowish or creamy-white. Avoid stark white or green colored undersides. With cut melons, select red, juicy flesh with black seeds. Keep at room temperature or refrigerate. Peak supply is May to August.

Apples

Over 200 million bushels of apples go to the market nationwide every year. Apples are of excellent quality and a large variety is available year round. In fall, over twenty varieties are in good distribution. The most popular varieties are McIntosh, Cortlands, Red Delicious, Yellow Delicious or Golden Delicious, Granny Smith, Rome, Empire, Macoums, and Wealthies. A new variety called Mutsu is a large, greenish fruit with a red blush. It is fine for pies and for decorating.

Apples are one of the most prized table fruits. An apple contains about 10 percent carbohydrates, mostly glucose and fructose that are easy to assimilate by the body. Vitamins C, B, iron, potassium, and phosphorus are also present in apples. So, the apple's reputation as a healthy fruit is well founded. Gastronomically, the apple is a most versatile fruit. From the decorative point of view, the apple is quite adaptable. The following decorations are easy to make.

The Apple Basket

Tool

a medium size paring knife

1. Cut out four wedges from a large apple, leaving a double handle and the lower half uncut.

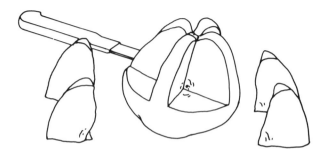

2. Carefully remove the apple pieces around the handles, including the core.

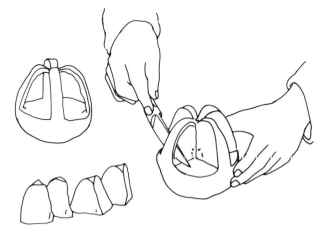

Use the apple basket as a garnish filled with citrus fruit sections.

The Apple Bird of Paradise

Tool

a medium or large paring knife

1. Slice off one third of the apple to be used as the bird's neck and head as explained in Step 2. This also provides a flat base for the bird. Next, cut out a small wedge from the top of the apple. Continue cutting larger wedges until the core is reached.

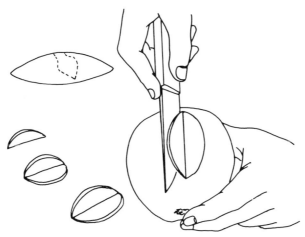

2. Using the one third of the apple set aside, cut a V for the head with the neck extending as shown.

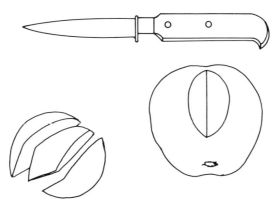

3. Continue cutting more wedges from the sides of the apple.

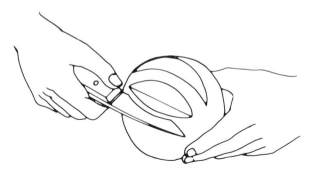

4. Assemble the sliced wedges, lapping them as shown, to form the tail on top and wings on the sides.

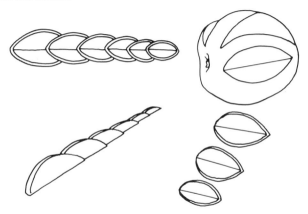

5. Insert two cloves for the eyes and attach the bird's head between the wings using a toothpick.

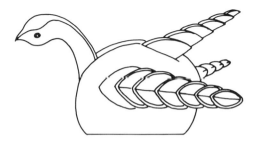

Gather several birds made of different colored apples and combine with fruit displays, melon baskets or any fruit arrangement. The bird of paradise is one of the most impressive fruit carvings. It is a very distinguished decoration when properly carved and displayed.

The Sawtooth Apple

Tool

    a medium size paring knife

    1. At the middle of the apple insert the knife to the core, alternating angles to obtain a sawtooth cut.

    2. Separate the apple halves, and remove the core.

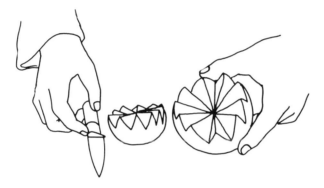

Use sawtooth apples baked as a garnish for meat, or serve as dessert.

The Apple Tulip

Tools

> a medium size paring knife
> a Parisian scoop

1. Starting at the blossom end of the apple, carve a thin slice down to ⅔ the thickness of the fruit.

2. Continue carving more slices around the circumference of the apple.

3. Remove the inside, first by carefully cutting pieces with the knife, then with the Parisian scoop.

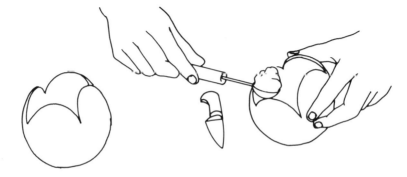

Fill the tulip cup with apple sauce, or any other stewed fruits.

Two Apples in One

Tools

>   an apple corer
>   a 12-section apple cutter
>   a small paring knife

1. Core two apples of identical size and shape, but of different colors such as red and yellow, or green and red.

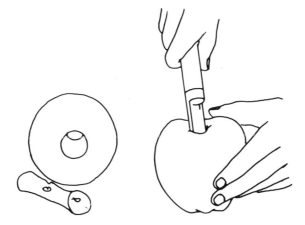

2. Press the apple cutter into the center of apple, down to $\frac{2}{3}$ its depth. Do the same with the other apple.

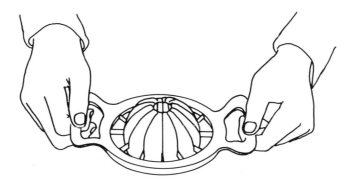

3. Remove the cutter. Slice off and interchange every other wedge of the two apples.

4. Assemble the cut out wedges as shown, giving a colorful contrast.

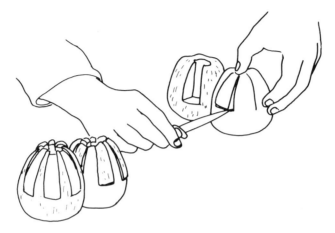

Use this decoration for children's parties or combine with a fruit arrangement.

V-Shaped Apple Slices

Tools

    an apple corer
    a medium size paring knife

1. Core a clean, unpeeled apple. Cut it in half and slice it into thin slices.

2. Fold slices in half as shown; avoid breaking them.

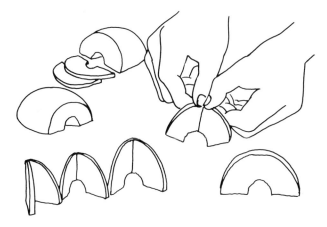

Surround apple pies, apple sauce, or other apple and fruit desserts with the V-shaped slices. Use different colored apples for contrast.

Avocados

The highly nutritious avocado is a fruit with an elongated pear shape and a yellowish green pulp. The chemical composition of the avocado is 60 percent water, which is low compared to other fruits. Lipids are present in large amounts, 30 percent; and carbohydrates, 3 to 8 percent. Avocado contains large amounts of vitamins A, B, B2, C, and others.

As food decoration, the avocado is a colorful garnish in salad. It should be peeled, sliced, or cut into cubes. To avoid discoloration, sprinkle lemon juice on the avocado.

The avocado can also be used as a container. Cut it in halves, do not peel. Remove the pit and garnish with

- jellied consommé garnished with sour cream and caviar,
- seafood salad,
- Waldorf salad.

Citrus Fruits

The citrus fruits best for food decorating are the sweet orange, grapefruit, lemons and limes. A characteristic of these fruits is their high content of vitamin C.

Sweet Orange

The sweet orange is the most popular of citrus fruits. It was first cultivated by the Chinese. A large variety of oranges is available on the U.S. market year round. Oranges are extensively grown in California and Florida. The cultivars of sweet orange are now very numerous. Notable among them are Valencia, Washington, Navel; and hybrids such as Temple (sweet orange and mandarine), and Tangelo (mandarine and grapefruit).

Grapefruit

Another popular species of the citrus family that originated in the West Indies is the grapefruit. There are two basic species available all year round: the white variety and the sweet, pink variety.

Lemon

The lemon is a versatile citrus fruit widely used in cooking. A few drops of lemon juice enhance the flavors of fish, poultry, and many desserts. The acid of lemon juice prevents cut fruit from turning brown when exposed to air.

Lime

This small, thin-skinned fruit of Indian origin is used as a lemon substitute. The lime with its delicate green color is a decorative fruit often used in combination with fish, salads, and desserts.

Grooved Lemon Slices

Tools

> a lemon peeler
> a medium size paring knife

1. With the peeler, cut V-shaped grooves lengthwise, spacing all grooves equally around the lemon.

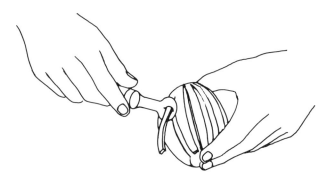

2. Cut the lemon in half lengthwise, then slice each half into $\frac{1}{8}$-inch thick slices.

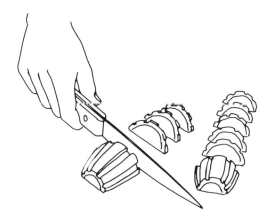

Use grooved lemon slices as a garnish for fish dishes or platters.

The Lemon Basket

Tool

    a medium size paring knife

  1. For the handle, make two parallel cuts, $\frac{1}{4}$ inch wide, half way through the lemon as shown.

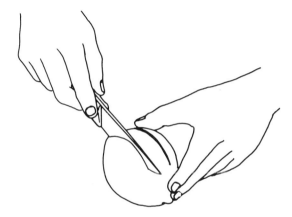

  2. Make two horizontal cuts to remove the two wedges around the handle. Make sure not to cut into the handle.

  3. Slice off the pulp around the handle.

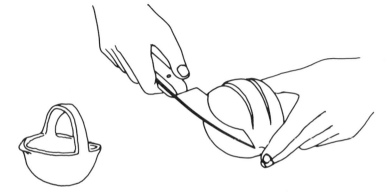

Fill baskets with parsley and use them to garnish fish dishes.

Lemon and Orange Wedges

Tools

    a large paring knife
    a small spoon

1. Cut the lemon and orange in halves. Scoop out the pulp.

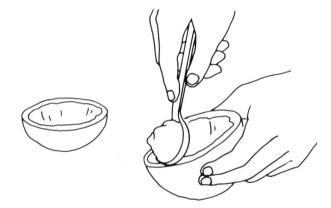

2. Fill the lemon shells with lemon-flavored gelatin, and the orange shells with orange-flavored gelatin.

3. Refrigerate until set; then cut into wedges.

Serve the wedges as dessert with fruit salad or use as decoration for gelatin desserts.

The Lemon Tulip Cup

Tools

> a medium size paring knife
> a Parisian scoop

Follow the same method for the Apple Tulip Cup. Using half a lemon shell remove the pulp without damaging the skin. Surround the lemon shell with a piece of foil to make a 1-inch-high collar. Fill the shell with fish mousse or lemon sherbet. Allow the filling to set in the refrigerator if using a filling containing gelatin or freeze the lemon shell if the filling is ice cream or sherbet.

Top left to right: Lemon tulip cup. (See the Apple Tulip Cup.) *Bottom left to right:* lemon basket; lemon half with twisted peel cut around the rim; decorative slices in the form of a bow and a cross; grooved lemon slices.

Grapes

Throughout the world, viticulture has increased so much that grapes are now one of the most abundant fruits. The cultivated varieties of grapes are very numerous and are divided into two groups: table dessert grapes and wine grapes. Both types are good for eating.

Grapes have digestive and therapeutic properties and are very nourishing. The percentage of carbohydrates is 18 to 20% in the form of glucose and fructose, both of which are easily assimilated. They also contain potassium, iron, sodium, calcium, magnesium, phosphorus, vitamins C and B complex.

California produces good varieties of red, green, and purple table grapes. In decorating food, grapes are often used in fruit salads and many combinations of fresh fruit presentations.

Sugar-Frosted Grapes

Here is a simple method for frosting grapes.

1. Dip a clean bunch of grapes in lightly beaten egg whites and roll in sugar. Shake off the excess of sugar and allow to dry.

Serve frosted grapes as dessert or use them in fruit displays.

Melon

Most melons such as cantaloupes, cranshaws, honeydews, honeyballs, Santa Claus, and Hand melons belong to the squash family. Another category, of which the watermelon is the best, belongs to the cucumber family.

Selecting a good melon that is not a "turnip" requires some training. Ripeness is an important characteristic. Melon does not ripen once it has been picked. It softens and rots, but does not become sweeter. A melon should not be soft when pressed, nor should it have bruises, dark spots, cracks, or shriveled skin.

Cantaloupe or netted melon with large pronounced netting over the skin surface indicates a superior melon. A creamy-yellowish background tone is a sign of ripeness. The Rocky Ford cantaloupes are among the favorites in August.

Hand melons, named after Mr. Hand, are grown in New York State and Vermont. They look like cantaloupes outwardly, but are juicier and more flavorful. They spoil quickly due to their high water content.

Cranshaws are ripe when they are tender at the blossom end and have a bright buttercup color. In the fall and later in the season, they turn medium green when ripe.

A ripe honeydew melon has a buttery or dull-matte finish, and gives off a melon aroma. A white, smooth waxy honeydew is unripe.

There are three types of watermelons

1. Relatively small fruit, about 6 to 10 pounds, spherical with dark green skin. They are called Sugar Baby, Summer Festival, and other names.
2. Larger spherical fruit, about 20 to 40 pounds, with uniformly colored skin, and a red pulp. These are the Dorias and the Florida Giants that can weigh over 40 pounds. The melons make great show pieces for buffets when carved and filled with fruit of the season.
3. Another popular type, varying in shape and shades with dark color on a light background is known as the Blue Ribbon, the Dixie Queen and the Klondike Striped.

Another variety of watermelon that contains less seeds has recently appeared on the market.

A good watermelon has a resonant sound when tapped.

The Honeydew Basket

Tool

a large paring knife

1. Select a medium to large honeydew. Cut a flat slice off the base of the melon so that the basket will rest upright. Cut a 2-inch-wide handle half way down the middle of the melon.

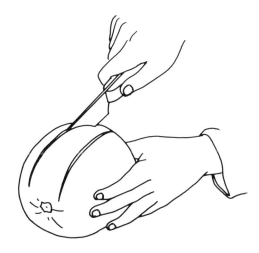

2. Make two horizontal cuts to remove the two wedges around the handle. Make sure not to cut into the handle.

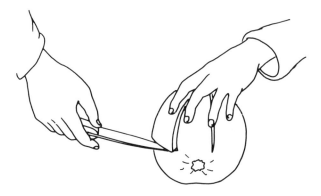

3. Cut a serrated edge all around the rim and handle of the melon.

4. Discard the seeds then fill the basket with fruit and melon balls extracted from the two wedges.

FRUIT DECORATIONS   **27**

🌸 The Melon Cradle

Tool

   a large paring knife

   1. Cut a flat slice off the melon as shown, so that it will sit at an angle.

   2. Outline the next cut with a felt pen in order to remove an oval shape slice. It should be about ⅓ of the melon.

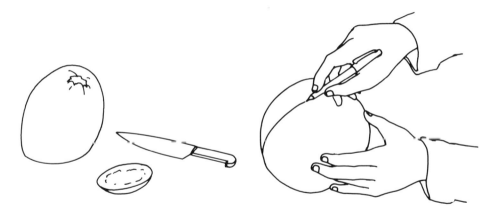

   3. Cut a serrated edge around the rim of the melon; remove the seeds and fill with melon balls or other fruit.

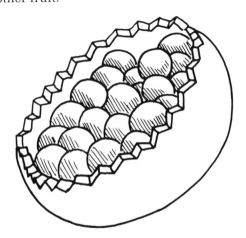

The Honeydew Melon Swan

Tools

> a paring knife
> a Parisian scoop
> a lemon peeler

Select a large honeydew melon.

1. Draw the front and the back of the swan on paper as shown. The upper line is the top of the melon. The lower line, the bottom. The dimensions of the drawings should be about ⅔ of the height of the melon.

2. Trace the patterns on the surface of the melon; first the front of the swan, then the back.

3. Carve the neck of the swan.

4. Carve the wing of the swan as shown.

FRUIT DECORATIONS 29

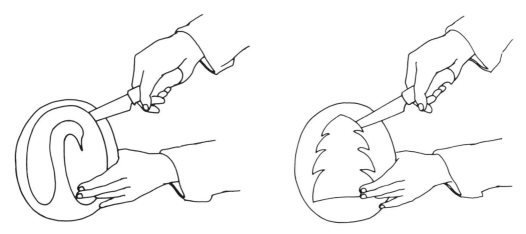

5. Scoop out the melon inside the swan.

6. With the cut out pieces make melon balls and put them back inside the carving. For color contrast, add melon balls from cantaloupe, cranshaw melon, or watermelon.

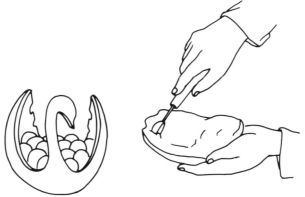

7. Decorate the wings of the swan using a lemon peeler. Chill the melon swan. The shell can be refrigerated and reused.

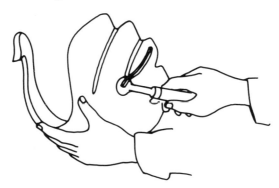

Papaya and Blueberries

Papaya is an intriguing and highly nutritious fruit. The following serving ideas are simple and colorful.

To serve a papaya, slice it across or lengthwise. Scoop out the seeds, and fill the containers with fruit of contrasting color such as blueberries, strawberries, or other fresh berries.

Gelatin desserts can also be poured into the papaya halves and refrigerated until set.

Lemon, orange, or grapefruit sherbets made individually or mixed can be scooped in the papaya shells.

Papaya halves can be cut across and lengthwise and filled with blueberries.

Pears

There are more than five thousand varieties of pears found throughout the world. Commercially, only a few varieties are cultivated because of their resistance to disease. Several varieties are available year round.

A ripe pear has juicy white flesh with a slight sweet-acid taste. Considerable amounts of the pears grown are used for canning for the food industry. In food decorating, pears are often used in combination with other fruit.

 Pear Mice

This is one of the most simple and charming ways of presenting a pear half.

1. Insert two sliced almonds in the stem end of the pear for the ears.

2. Stick two small chocolate chips for the eyes, and a larger chocolate chip for the nose.

3. Add a lemon twist for the tail, and the mouse appears!

Serve pear mice as dessert, covered with fudge sauce, or arrange them on top of ice cream. They are great fun for children's birthdays.

Pineapple

The pineapple, grown extensively in Hawaii, has firm whitish yellow pulp and a very delicate flavor. It is available year round and large quantities are used in canning. Nutritionally, pineapple contains 15% sugar, and malic and citric acids. In food decorating, the shell and leaves of the pineapple can be used for various dessert presentations.

 # The Pineapple Bird

Tool

a large paring knife

1. Select a ripe pineapple. Cut it in half vertically, leaving all the leaves attached to the bottom half.

2. Carve the head of a bird from the core of the pineapple half that has no leaves. Secure it with a toothpick as shown. Insert two cloves for the eyes.

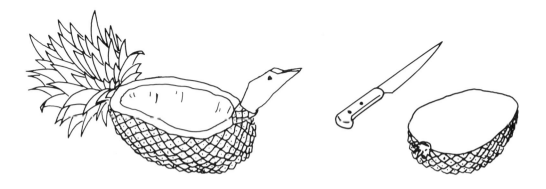

3. Fill the pineapple with the remaining pineapple chunks and other fruits.

Use the pineapple bird to serve fruit, sherbet, or ice cream.

The Pineapple Cart

Tool

a large paring knife

1. Make two parallel cuts across the pineapple leaving both ends uncut, as shown.

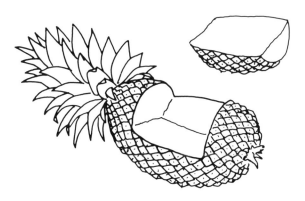

2. Slice two ½-inch-thick round slices from another pineapple. Attach them with toothpicks to imitate the wheels of the cart. Decorate the wheels with pear wedges for spokes and a black cherry in the center.

The pineapple cart filled with fruit makes an unusual display piece for a buffet.

The pineapple can also be cut in half horizontally and filled with fruit.

3
Vegetable Decorations

Here is a guide to selection and care of fresh vegetables used for decorating food.

- **Carrots** Look for firm, well shaped roots with a good orange color.
- **Cucumbers** Choose medium sizes with good green color. Avoid very large, puffy ones or any that are yellow.
- **Leeks** Select well-blanched bunches (white coloring extending 2" to 3" from bulb base). Choose small or medium leeks for the most tender eating. Refrigerate in a plastic bag and use within three to five days.
- **Lettuce, iceberg** Heads should be firm but should give slightly when squeezed. Core the iceberg by pressing it core-end down on the kitchen counter. Twist to remove the core and rinse end up under cold running water. Drain thoroughly in a colander or rack with cored-end down. Store in a tightly closed plastic bag or lettuce crisper.
- **Mushrooms** May be white, tan, or cream-colored. The freshest mushrooms are closed around the stem by a thin veil. However, those having open veils caused by loss of water are just as nutritious but have a more pungent flavor. Refrigerate and cover with a damp paper towel to aid in moisture retention. Avoid storing in plastic bags; this hampers air circulation.
- **Onions, green** Select young and tender bunches with fresh green tops. Keep refrigerated in a plastic bag and use as soon as possible.
- **Peppers** Should be fresh, firm, thick-fleshed with bright green coloring which may be tinged with red. Immature peppers are usually soft and dull-looking.
- **Potatoes** Choose firm, clean, and relatively smooth ones free of cuts and bruises. Avoid green potatoes and those with sprouts. Never refrigerate potatoes but keep them in a cool, well ventilated, dark area. Keep away from light, which can cause greening, and moisture.
- **Radishes** A variety of sizes and shapes is available. All should be fresh, smooth, and well formed with few cuts or pits. Avoid spongy radishes.
- **Rutabagas** Should be firm, heavy for their size, smooth, and not cut or punctured. Size is not a quality factor.
- **Squash, summer or winter** See the section on squash for detailed information.

- **Tomatoes** Choose smooth, firm, and plump tomatoes with good color. Most tomatoes require further ripening. When red-ripe, refrigerate and use within a few days.
- **Turnips** Select firm, smooth, and medium size turnips. Avoid yellowed or wilted tops which indicate old age.

Carrots

The carrot is one of the most important garden vegetables in the United States. Many varieties are available throughout the year.

Carrots are used for cocktails, appetizers, hors-d'oeuvres, vegetables, salads, desserts, and they are a base for many sauces. Like most vegetables, carrot is low in protein and fat but rich in vitamins, especially A, B, E, and carbohydrates.

The typical color comes from carotene, a good source of vitamins. Raw carrot, cut into sticks, has become along with celery the most popular crunchy appetizer in America.

The following decorations can be used as edible appetizers or centerpieces on cold food dishes. Combined with other vegetable flower arrangements, the carrot spider mum, centered with a small ripe olive or a tiny piece of green pepper, appears as a delicate, colorful, and intriguing flower.

The palm tree made with a crisp carrot for the trunk and topped with a carved green pepper for leaves is strikingly effective and easy to do.

To obtain the best possible results when carving carrots, it is imperative to select the freshest and the largest, with a bright orange color and a tender heart. If the carrots lack the crispness required for carving, caused by insufficient moisture, soak them in ice cold water for one hour.

Flower arrangements and any other decorations made of carrots are at their best if used as soon as prepared. They have a tendency to dry out at room temperature, but will retain their freshness when misted with water the same way real flowers are treated. If the decorated arrangements are to be used days after they are prepared, they will keep fresh and crisp under refrigeration and in water but will lose some of their nutritive value. An elaborate display of carrot flowers and carvings may be done days ahead, in which case, the water used to store the vegetables is a good source of vitamins and should be saved.

Carrot Balls

Tool

a Parisian scoop smaller than ½-inch in diameter

Using either a raw or a cooked carrot, press the Parisian scoop firmly into carrot and extract a round ball. Continue the same procedure using the entire carrot.

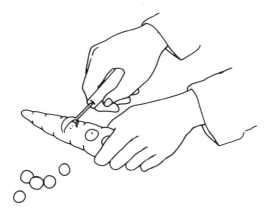

Arrange carrot balls in rows on hot or cold foods. Use them cooked as garnish in consommés or clear soups. They can also be used with white turnip balls.

The Carrot Daisy

Tools

>an apple cutter and corer with 12 sections
>a 2½-inch paring knife

1. Cut a section of a peeled carrot 2 inches long and 1 inch in diameter. Place the center of cutter exactly over the carrot heart and press firmly down about 1½ inches.

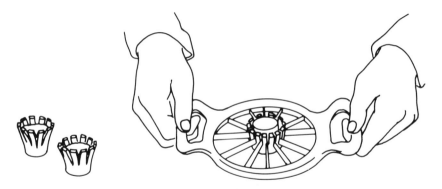

2. Remove the cutter carefully. Cut off carrot core with a paring knife.

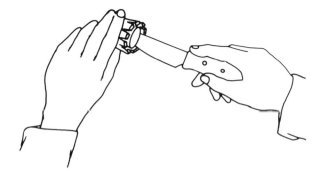

3. Garnish the center of the daisy with a parsley sprig, cauliflower flowerettes, or half of a cherry tomato.

Use the carrot daisy for appetizers or decorations for salads and cold meat platters.

The Carrot Flower Cup

Tools

    a lemon peeler
    a 2½-inch paring knife

1. Wash and peel a medium to large carrot. Cut V-shaped grooves lengthwise with the lemon peeler, spacing all grooves equally around the carrot.

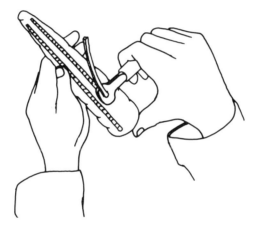

2. Trim smaller end into a cone. Using the cone surface as a guide, carve two very thin cups around the circumference of carrot.

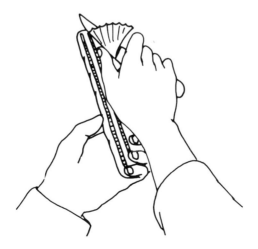

3. Put the two cups together to form one as shown.

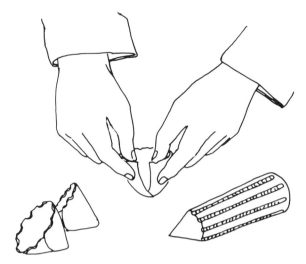

4. Garnish the center of the flower cup with a tiny radish ball with the red skin showing. Refrigerate in cold water.

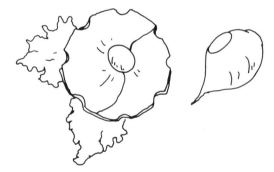

Use carrot flower cups as appetizers or garnishes for salads or cold platters.

The Carrot Lily

Tool

    a 2½-inch sharp paring knife

1. Wash and peel the carrot. Starting at the tapered end, cut a 2-inch petal shaped sliver leaving it attached to the base of the carrot. Repeat the same procedure around the circumference of the carrot three to four times.

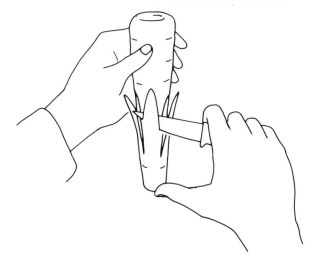

2. Break off the base of carrot to obtain a flower cup.

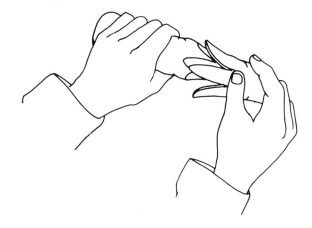

3. Garnish the center of the lily with a dab of cream cheese, a sprig of parsley, a small ripe olive, or a slice of stuffed green olive.

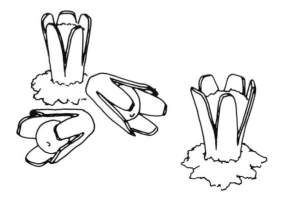

Use carrot lilies as appetizers or garnishes for cold platters.

Carrot Pinwheels

Tools

a lemon peeler
a 4½-inch paring knife

1. Wash and peel a thick carrot. Cut V-shaped grooves lengthwise with the lemon peeler, spacing all grooves equally all around the carrot.

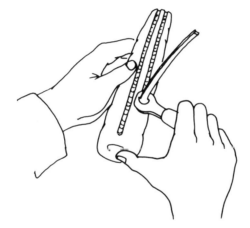

2. Slice the carrot into ⅛-inch-thick slices.

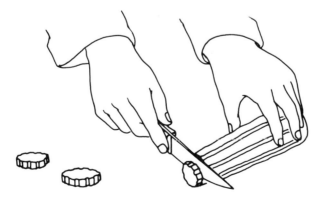

Use carrot pinwheels raw in salads or cooked as a garnish for soups and stews.

Carrot Stick Bundles

Tool

    a medium French knife

1. Peel and slice a red onion into rings ¼-inch thick.

2. Insert carrot sticks into small onion rings to form tight bundles. Place them in ice cold water for a half hour to crisp.

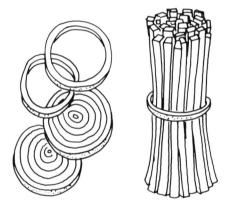

Use carrot stick bundles as appetizers and garnishes.

Corrugated Carrot Slices

Tool

a corrugated vegetable slicer

1. Wash and peel a thick carrot.

2. Using the slicer, cut ¼-inch-thick diagonal slices.

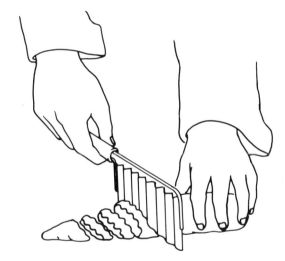

Use corrugated carrot slices in salads or as a cooked vegetable in meat stews.

The Palm Tree

Tools

a 2½-inch paring knife
2 plain wooden toothpicks

1. Wash and peel a 6-inch-long carrot. Start at the thickest part of the trunk making a ½-inch-deep slanted cut into the carrot.

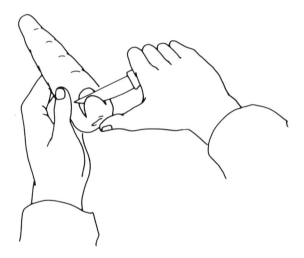

2. Remove a sliver directly above the first cut.

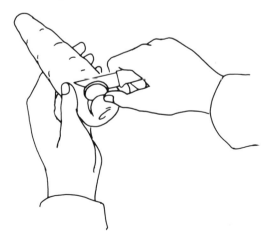

3. Continue in the same manner over the whole surface of the carrot.

4. Carve a green pepper into sawtooth pieces.

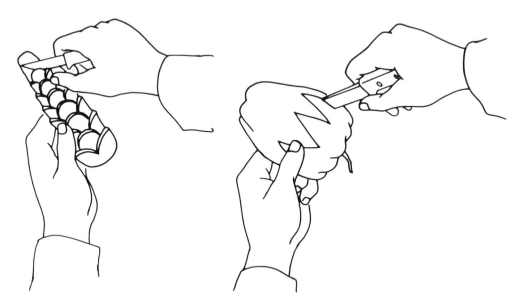

5. Secure carrot on a turnip base with toothpicks. Crown with top part of the green pepper half.

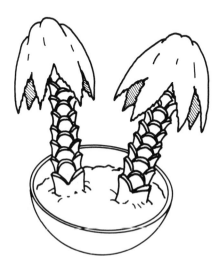

Use the palm tree for buffet decorations. A better effect is obtained using two palm trees pointing in different directions.

The Spider Mum

Tools

    an electric or manual slicer
    a 2½-inch sharp paring knife
    a plain or orange toothpick

1. Wash, peel, and trim ends of a 5- to 6-inch-long carrot. Slice lengthwise into ⅛-inch-thick slices. Five slices are needed for best results.

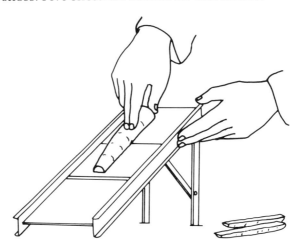

2. Make cuts ¼ inch apart through the length of carrot slices. Note that two slices are cut free from thickest end of carrot.

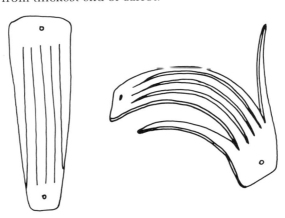

3. Insert a toothpick into the ends of carrot slices.

4. Arrange curled slices into an attractive, five-petaled flower. Conceal tip of toothpick with a tiny piece of ripe olive.

Use the spider mum on cold food displays.

Cucumbers

The cucumber is not new on our planet; it has been known in the Orient for over three thousand years and has always been a very popular food. Although the cucumber lacks nutritive value due to its high percentage of water (over 90%), it is a most refreshing food and contains some vitamin C, thiamine, and riboflavin.

All of the following decorations are easy to create.

 ## The Cucumber Basket

Tools

a paring knife
a Parisian scoop

1. Cut a 2-inch section of a whole cucumber. Slice off a shaving so the basket rests flat. For the handle make two vertical cuts about halfway through the thickness of the cucumber.

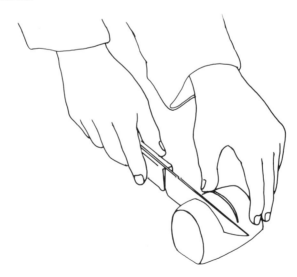

2. Make two horizontal cuts to remove two wedges without cutting through the handle. Scoop out the inside, including the handle of the cucumber basket. Soak in ice cold water.

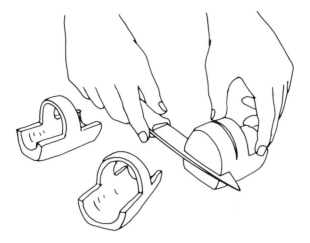

The cucumber basket is very versatile. It can be garnished with celery and carrot sticks, tiny white or green asparagus spears or marinated green beans.

The Cucumber Chain

Tools

    a large paring knife
    a round cutter

1. Slice the cucumber into rings ⅜-inch thick. Core each slice with the round cutter.

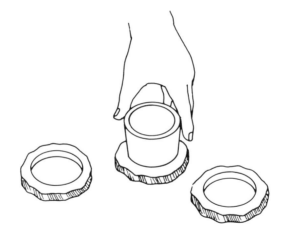

2. Slit every other ring and link the chain to desired length.

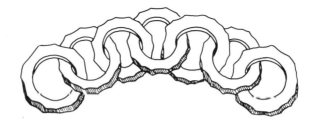

Cucumber Fish Scales

Tools

    a paring knife
    a slicer

1. Slice a medium size cucumber very thinly. Cut slices in half.

2. Arrange the cucumber slices over a whole cooked fish or fillet to imitate the scales.

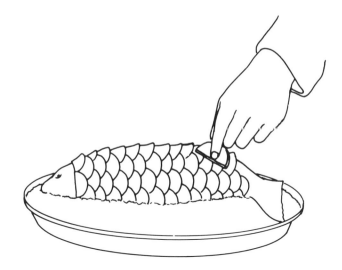

The Cucumber Tulip Cup

Tools

a paring knife
a Parisian scoop

1. Make a slanted cut 2 inches from the end of the cucumber. Make three more cuts around the circumference of the cucumber.

2. Break off the cup and scoop out the inside. Cut the end of the cucumber flat and continue with additional cups.

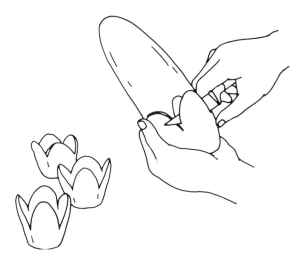

Use cucumber cups to hold cold sauces such as horseradish, mayonnaise, and tartar. They can also hold capers, chopped egg yolks, chopped egg whites, or chopped onions.

Left to right; top to bottom: Radish daisies and a lemon basket. Coconut souffles served in coconut shells. The apple bird of paradise. Versatile carrot decorations. Orange souffles in orange halves.

Left to right; top to bottom: A bouquet of radish daisies. A salmon picture plate. Honeydew melon swans and melon baskets. Cucumber creations. Butter mold decorations.

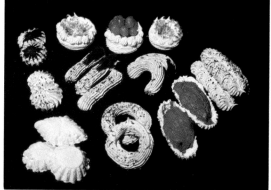

Left to right; top to bottom: Meringue mushrooms. Meringue fancies. Meringue swans. A meringue basket filled with raspberries. Butternut squash vase with turnip daisies.

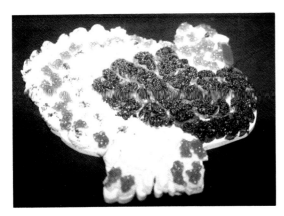

Left to right; top to bottom: Thanksgiving turkey appetizer tray. Fish and seaweed cut from color sheets. 100-year-old eggs, snowman, and ladybug decorations. A woven bread basket. Watermelon whale.

The Cucumber Viking Ship

Tools

    a paring knife
    a 2-inch round cutter
    a 2¼-inch round cutter
    a Parisian scoop
    a Bamboo skewer
    10 toothpicks

1. Cut off one third of the cucumber horizontally. Scoop out the inside with a Parisian scoop.

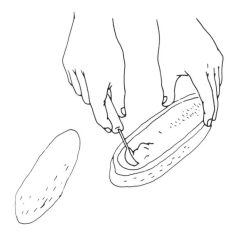

2. Shape the two sails with the round cutters as shown. The flag is cut with the leftover piece of cucumber.

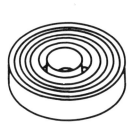
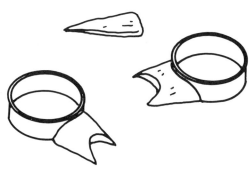

3. Insert the bamboo skewer through the two sails and the flag.

4. Set the sails at the end of the cucumber as shown.

5. Peel and slice a large carrot diagonally into 10 thin slices. Insert into the toothpicks and set around the ship to make 10 oars.

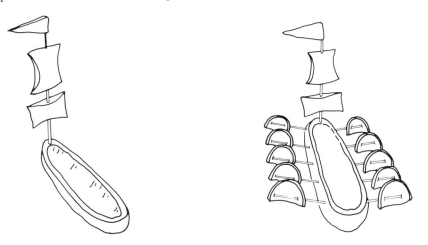

The cucumber Viking boat is an ideal container for cold sauces to accompany fish or for salad dressings.

More Ideas Using Cucumbers

1. For appetizers serve thin slices of crisp cucumber topped with smoked salmon roulades stuffed with cream cheese.
2. Cut a 5- to 6-inch cylinder from a large cucumber. Remove the core with a paring knife. Stuff the cavity of the cucumber with a firm ham spread. Refrigerate and slice into ¼-inch rings. Serve as appetizers.

Belgian Endive

The Belgian endive, also known as witloof, is much sought after by gourmets for use in salads, braised dishes, and other preparations. The elongated white leaves with yellow tips are best used as containers or boats in food decoration.

Endive leaves make good containers for caviar garnishes. Place the caviar surrounded with a ring of ice in a suitable bowl. Surround the caviar with endive leaves containing chopped egg yolks, chopped egg whites, capers, and lemon pieces. Put two garnishes in each endive leaf. This particular method of serving the garnish for caviar is practical and original, and quite decorative.

Crisp endive leaves can also be used to hold shrimp or seafood salads. Another way to present endive is to spread the leaves of a whole endive to make it "bloom" into a flower.

Leeks

In the United States, the leek is not as popular a vegetable as it is in Europe. The leek is a vegetable for fall, winter, and spring consumption. It is an excellent ingredient for soups. It is also very good as an appetizer when made into a quiche, or braised and served with a bouquet of cooked vegetables.

The leek has modest nutritive value with 2 percent protein, and 7 percent carbohydrates, but it is good from the dietetic point of view, mostly for its antiseptic and diuretic properties.

In food decorating, the white part of the leek can be carved into an attractive flower. The green part of the leek is an essential element for fine decorations and is chiefly used to imitate leaves and flower stem arrangements.

 ## The Leek Blossom

Tool

a medium size paring knife

1. Trim the root off the leek, then cut a 2- to 3-inch section from the white part of the leek.

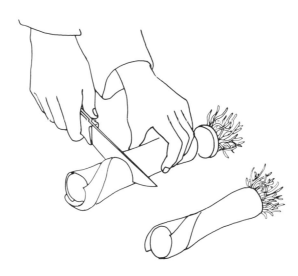

VEGETABLE DECORATIONS   59

2. Slice ⅛-inch-wide parallel strips through the center of the leek, leaving about ½ inch of the root end uncut. Continue the same procedure all around the leek.

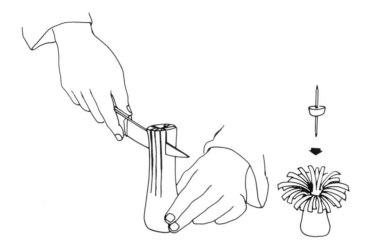

3. Open the cut leek without breaking the strips. Stick a small, round piece of carrot onto a plain toothpick and push into the root end. Store the flower in cold water.

More Decorations Using Leeks

The green part of the leek is often used as stems and leaves for floral decorations. The green leaves must be blanched for 2 to 3 minutes, then cooled in ice cold water. The leek becomes pliable and easy to carve. Some designs created with the stems of green leeks are shown in the illustration. The stems of chives can also be valuable for decorating food.

Field Mushrooms

Field mushrooms are abundant and available year round. Pennsylvania is one of the major centers of mushroom production. The cultivated field mushroom is harvested young, either as a button or as a cup when the cap has partially opened.

Mushrooms can be eaten raw or marinated. They are often cooked and are used in a large variety of dishes.

The decorative value of the field mushroom is limited. The buttons can be used to boost dull potato and green bean salads. Mushroom cups can be turned into fluted mushrooms. Cooked, they make a fine decorative garnish for fish and meats.

Fluted Mushrooms

Tool

a small paring knife

1. Starting at the top center of the mushroom remove a fine peeling of the mushroom, following a contour with the knife as shown. Continue the same procedure all around the mushroom to obtain fine evenly spaced grooves.

2. To make a star on top of the mushroom, press the tip of the knife into the center of the fluted mushroom.

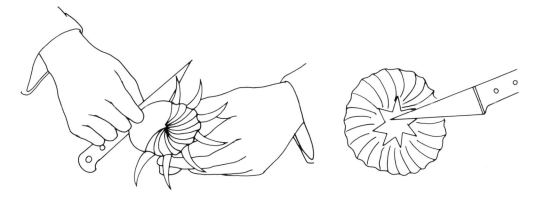

Cook the fluted mushrooms in lemon juice and butter, and use to garnish fish and meat.

Olives

The olive is the fruit of an evergreen tree. The cultivated olive is rich in oil. The pressed, crushed olives produce virgin oil which is the most prized. In practice the various grades of olive oil are classified on the basis of the percentage of acidity (oleic acid) and range from extra virgin to superfine virgin to fine olive to simply virgin olive oil. Inferior quality oils are named *olive oil*. Besides the olives used for the extraction of oil there are the table olives. These firm olives are picked green or ripe (black) and are preserved with or without pits, or stuffed. California produces excellent quality table olives that are adaptable in food decorating. They range in size from tiny to giant.

The Olive Chain

Tool

a small paring knife

1. For best results, select large to jumbo ripe olives. Slice them lengthwise into small strips and arrange into a chain.

Use olive chains to decorate potato salads and cole slaw. Stuffed green olives can be sliced and used as border designs with salads.

Parsley

Parsley has become almost indispensable in our kitchens. It is used constantly in everyday cooking. It is rich in minerals, iron, and vitamin C, and it is recognized for its appetite stimulating qualities.

In decorating food, parsley is probably the most used herbaceous plant. Many simple dishes are garnished with chopped parsley, sprigs, or bouquets. The curly leaves of fresh parsley have a beautiful bright green color.

It is interesting to note that parsley sprigs can be deep fried. The parsley should be clean and dry. Fry in a deep fryer at 360°F for less than a minute. Fried parsley should be green and crisp. It is used as garnish with fish and is very tasty.

Peppers

Two species of peppers are recognized in the culinary world: the sweet table peppers, mostly green, red or yellow, and the seasoning varieties which are negligible in food decorating. The sizes and shapes of sweet peppers are innumerable. In the United States, the leading varieties are the Bell Boy, California Wonder, and the Merrimack Wonder, among others. Nutritionally, sweet peppers are low in calories and contain vitamins A, B, C, and E.

From a decorative standpoint, peppers are useful due to their shapes and bright colors.

 ## Pepper Baskets

Tool

    a medium size paring knife

1. To carve the handle, make two parallel cuts $\frac{1}{4}$-inch wide, halfway down the green pepper. Also, make two horizontal cuts to remove the two wedges around the handle. Do not cut through the handle. Remove the seeds.

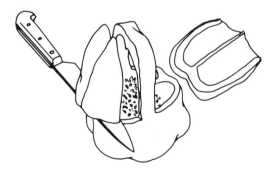

Fill pepper baskets with olives, baby corn ears, or dips for appetizers. Use green and red peppers for color contrasts.

Pickles

In the United States, kosher dill pickles are commonly used as condiments with cold meats and salads. They are sold whole or sliced. Gherkins are small cucumbers that are grown exclusively for pickling. These two varieties of pickles are the most adaptable for food decorations. They can be turned into fans and other simple designs.

 ## Pickle Rings

Tool

a small paring knife

1. Select a medium size, firm dill pickle. Slice in half lengthwise.

2. Make a series of slices about $\frac{1}{8}$ inch apart down the length of one half, leaving a $\frac{1}{4}$ inch spine to hold the slices together.

3. Spread the pickle out and shape into a ring.

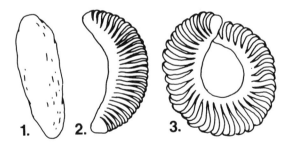

Use pickle rings as a garnish with cold meats and salads.

 Pickle Fans

Tool

a medium paring knife

1. Choose medium pickles for best results. They may be sliced in half lengthwise.

2. Make a series of slices about $\frac{1}{8}$ inch apart down the length of the pickle, leaving $\frac{1}{2}$ inch at the stem end uncut, in order to hold the slices together.

3. Spread the fan out on the working surface.

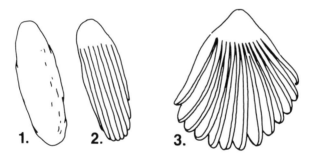

Serve pickle fans as a garnish with cold meats and salads.

Potatoes

The potato is one of the most widely known and used food plants. World production is over 350 million tons. In the United States, the potato is suitable for a large variety of dishes. The part of the potato used as food is called the tuber. It contains an average of 65 percent starch, 4 percent sugar, 9 percent protein, potassium, phosphorus, iron, and copper. The nutritional value of potatoes should not be overlooked. The equivalent of 3 to 4 ounces of bread corresponds to 1 pound of potatoes as far as the calories are concerned. Potatoes are also recommended for those suffering from high blood pressure because of their low percentage of salts. They also contain some vitamin C, thiamin (B1), and riboflavin (B2).

Potatoes used in food decorating are often deep fried. The following methods are relatively simple to create.

Potato Roses

Tool

a potato slicer

1. To make a potato rose, cut one large unpeeled potato and one small potato into paper thin slices; do not store the sliced potatoes in water. Trim a 3-inch-long carrot to the thickness of a pencil. Roll a small slice of potato around the carrot.

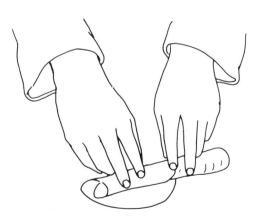

2. Roll additional small potato slices around the carrot, securing them in place with toothpicks.

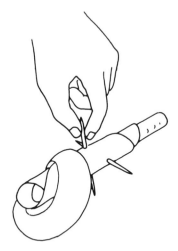

3. Use large slices of potato for the outer leaves of the potato rose. When the flower is fully formed, place it in cold water for 5 minutes. Drain and deep fry the potato rose until golden brown. Cool the flower then remove the piece of carrot and the toothpicks.

Use potato roses as appetizers and to garnish meat platters.

Fried Potato Nests

Tools

    nest baskets
    a potato slicer (mandoline)
    a deep fryer

1. Several sizes of the wire nest baskets are available. On the right, one set of baskets is shown open with the clamp. On the left, it is assembled with the small basket fitting into the large one and the clamp on the handle.

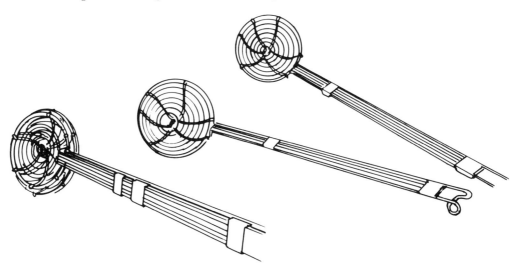

2. Slice *gaufrette* or wafer potatoes with the slicer. Arrange the potatoes into the large basket to form a nest. Do not wash the potato slices.

VEGETABLE DECORATIONS   69

3. Press the small basket against the potato slices in the large basket. Clamp the two baskets together at the handle.

4. Deep fry the potato nest in clean oil heated to 360°F until golden brown. Remove the clamp, lift the small basket, and tap the large basket against a hard surface to release the nest.

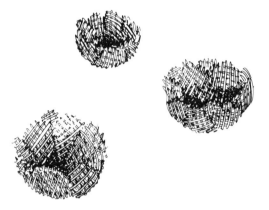

The same procedure is used with straw potatoes and gives a more realistic looking bird's nest.

Use potato baskets as garnishes for meat, filled with Parisian potatoes, puffed potatoes, etc. Fill them with quail eggs and serve as appetizers.

Radishes

Garden radishes are popular as food although their nutritive value is almost insignificant. The most common varieties are the red globe shaped variety and the long root white radish also known as the white icicle. Another variety is the oriental radish known as daikon. It grows to three feet in length and may weigh forty pounds or more. The smaller daikon, eight to ten inches in length, is preferable. Large ones are often spongy. Radishes make inexpensive decorative blossoms for garnishes, or appetizers.

 ## The Radish Daisy

Tools

>   a small paring knife
>   a 12-wedge fruit cutter

1. Cut a sliver off the top or the bottom of a red radish as a base. Place the radish between two bamboo skewers.

2. Press the cutter down the middle of the radish until it reaches the skewers.

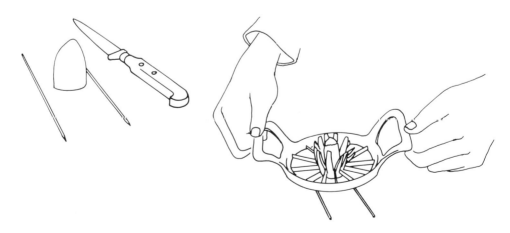

3. Remove center at the base. Push the daisy out of the cutter. Decorate the center with orange cheddar cheese balls.

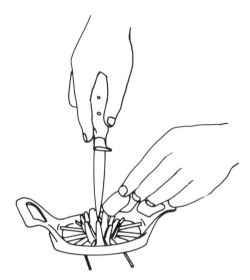

Use radish daisies as appetizers or as garnishes for salads or cold plates.

The Radish Fan

Tool

a small paring knife

1. Hold the radish stem end down. Make a diagonal cut at the bottom of the radish.

2. Make the next cut directly above the first to obtain a thin sliver attached to the radish. Continue with more cuts as high as the tip of the radish.

3. Turn the radish and continue on the opposite side with the same procedure as described in Steps 1 and 2. Place the fans in ice cold water to bloom.

The Radish Rose

Tool

an electric or manual slicer

1. Slice a red radish paper thin.

2. Overlap the slices and shape them into a rose as shown. Decorate center with a slice of stuffed green olive or a round sliver of radish.

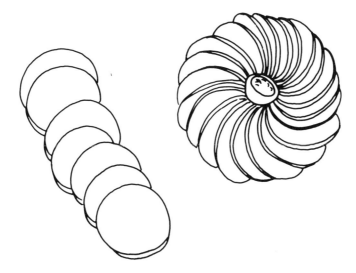

Use radish roses to decorate cold mousses, cold buffet plates, and platters.

The Radish Tulip

Tools

    a small paring knife
    a grapefruit knife

1. Hold a long radish stem end down. Make one thin slice from the top outermost edge of the radish, down, but not completely through.

2. Repeat Step 1 three more times around the circumference of the radish.

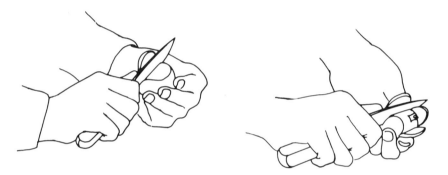

3. Remove the center at the core with a grapefruit knife. Store in ice water to bloom. The center can be saved and sliced for salads and sandwiches.

Use radish tulips as appetizers garnished with olives or cream cheese. They also make colorful decorations in salads and cold meat platters.

The Red Radish Blossom

Tool

a small paring knife

1. Hold a large round red radish stem end down. Make a slanted cut at the lower portion of the radish.

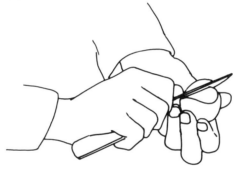

2. Make more cuts around the middle and the top of the radish at staggered intervals.

3. Slice another small red radish into very thin rings. Cut rings in half. Fit the thin rings between the wedges of the large radish.

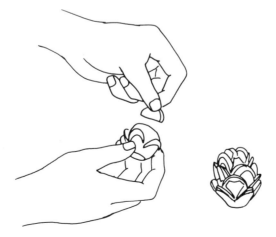

Use radish blossoms as garnishes with salads and cold food displays.

The White Radish Spray

Tools

a peeler
a small, sharp paring knife

1. Peel the radish. Holding the radish stem end down, make thin parallel slices lengthwise. Do not cut through the radish.

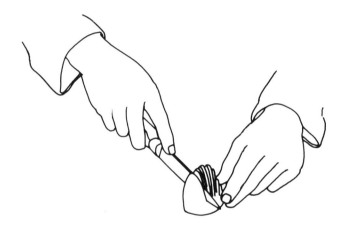

2. Turn the radish holding the strips together, then cut more slices as in Step 1 to obtain a crisscross pattern.

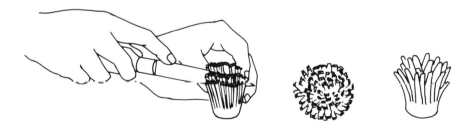

3. Chill the radish sprays in ice cold water. Red radishes can give the same results. Alternate white and red radish sprays for color contrasts.

Use radish sprays as decorations for sushi, cold meat and seafood platters, or serve them as appetizers.

Squash

The collection of squashes, pumpkins and other plants of the gourd family is profuse. According to the United Fresh Fruit and Vegetable Association, there is a squash for every season and every taste, so many varieties in different colors, shapes and sizes.

The hard-shelled squashes—the winter varieties include

- **Acorn** The acorn squash is wide-ribbed, slightly oval, and pointed at one end. The dark green shell changes to orange in storage.
- **Banana** The banana squash is large, long, cylindrical, and pointed at both ends. Its moderately thick, pale gray to creamy white shell can be smooth or somewhat wrinkled.
- **Buttercup** The buttercup squash is dark green with grayish spots and stripes. It is drum-like in appearance, about 4 to 5 inches long, and is crowned with a grayish turban-like top. It has sweet orange flesh.
- **Butternut** Cylindrically shaped with a thick neck, the butternut squash has a smooth creamy brown or dark yellow shell with deep orange flesh.
- **Hubbard** Large, about 10 to 16 inches long, and spherical, the hubbard squash is tapered at both ends. The shell is ridged, warted, and varies in color from dark blue, to gray, to orange red. The hubbard has sweet orange flesh.
- **Spaghetti Squash** The spaghetti squash has creamy white flesh, which is spaghetti-like when cooked. This variety is oval shaped with a light yellow smooth shell.
- **Turban** The turban squash is used only for decorative purposes. It is drum-shaped with a striped red-orange shell. As the name implies, it is turban-shaped.

The soft-shelled squashes—the summer varieties are

- **Cocozelle** This squash is cylindrical with alternating dark green and oval ribs. Choose squash about 6 to 8 inches long.
- **Chayotte** Round or pear-shaped, the rind of this squash can be smooth, ribbed, or covered with smooth white to dark green spines. Choose firm, hard squash.
- **Scallop** Also known as cymling and pattypan, this squash is small, flat, and pie-shaped with a scalloped edge. It can be white, yellow, light-green, or striped. Select small squash, 3 to 4 inches in diameter.
- **Yellow Crookneck** The yellow crookneck squash has bright yellow and lightly pebbled skins. It is 4 to 6 inches long with a curved neck.

- **Yellow Straightneck** This squash is similar to crooknecks with either a smooth or lightly pebbled skin.
- **Zucchini** Slender, dark green, and cylindrically shaped, zucchini have thin, smooth skins that are frequently striped with pale yellow or white. Choose thin squashes 4 to 9 inches long.

Many squashes are valuable for food decorations.

1. The butternut squash is a perfect vase for a bouquet of flowers. (See butternut squash vase in the color insert.)
2. The pumpkin becomes a fancy container when carved. It is a perfect soup tureen.
3. The yellow crookneck squash can be displayed on a buffet table and looks like a bird when sitting on its base.
4. The turban squash can be carved into a bowl and filled with salads or vegetables.
5. Zucchini can be cut into a sail boat like a cucumber boat.
6. A combination of gourds and squashes make a colorful centerpiece for a fall buffet.

Tomatoes

In total production tomatoes are now the nation's fifth ranking vegetable. They are available year-round.

The raw tomato has modest nutritive value with a water content of 90 percent and 4 to 5 percent carbohydrates. Vitamin A, thiamin and riboflavin are also present. Its bright red color helps stimulate the appetite and lends itself to a good number of decorative displays.

 ## Tomato Aspic Wedges

Tool

a large paring knife

1. Cut a medium size tomato in half. Scoop out the pulp and seeds.

2. Place the tomato halves on a plate and fill the shells with well seasoned tomato aspic.

3. Refrigerate until set and cut into wedges.

Serve tomato aspic wedges with salads, as a side dish with chicken or fish, or as decorations with cold buffet foods.

Tomato Mushrooms on a Cracker

Tools

a small paring knife
a pastry bag fitted with a plain round tube

1. Pipe out soft cream cheese on a cracker to form the stems of two mushrooms.

2. Cut two round caps from a cherry tomato.

3. Place the caps on top of the cream cheese and decorate tomatoes with cream cheese dots using a paper cone.

Use tomato mushrooms as appetizers or as garnishes with cold dishes.

The Tomato Poinsettia

Tools

    a small paring knife
    a small spoon

1. Select a ripe firm tomato. Place it stem end down on the working surface. Score the top of the tomato down to ¾ inch from bottom. Make another cut to form a triangle shaped petal. Continue all around the circumference of the tomato.

2. Peel back the petals.

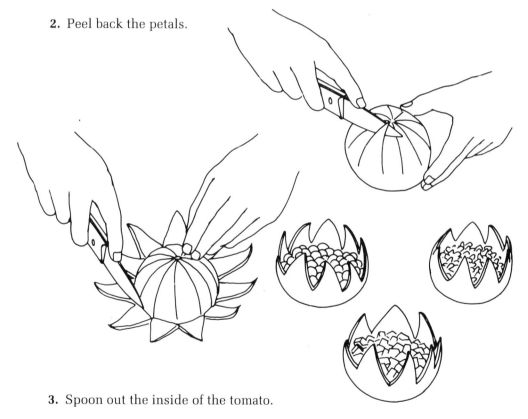

3. Spoon out the inside of the tomato.

4. Garnish the center with vegetables of contrasting colors.

The tomato poinsettias make ideal containers for individual servings of vegetables, meat, or fish salads.

The Tomato Rose

Tool

    a medium size paring knife

  1. Select a medium size, firm red tomato. Peel into a continuous 1-inch strip of thin skin, using a sawing motion with the knife.

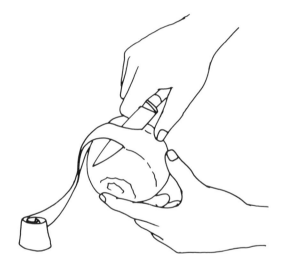

  2. Roll the tomato skin strip into a tight ring to form a rose.

3. Display the tomato rose with parsley stem.

The tomato rose is an attractive garnish for a great variety of food platters.

4
Sheets of Color

One of the most innovative methods of decorating food has taken root in my kitchen. The color sheets or aspic sheets, as they are called, are the most up-to-date approach to decorating food. Artistically inclined chefs and food decorators are relying on color sheets to embellish their award-winning food displays presented to panels of culinary experts gathered at food shows across the country.

Color Sheets

The method of preparation of color sheets is simple.

1. Combine 1 oz (25 g) of unflavored gelatin with $\frac{3}{4}$ cup (2 dl) of water. Allow the gelatin to soften, then gently heat the mixture over a double boiler.

2. Depending on the color desired, the following ingredients are pureed in an electric blender with the gelatin mixture.

- For an orange sheet, add $3\frac{1}{2}$ oz (100 g) canned or freshly cooked pimientos.
- For a red sheet, add 2 oz (60 g) pimientos and 1 oz (30 g) tomato paste.
- For a yellow sheet, add 4 hard-boiled egg yolks.
- For a white sheet, add 3 hard-boiled egg whites.
- For a green sheet, add 3 oz (75 g) cooked spinach.
- For a black sheet, add $3\frac{1}{2}$ oz (100 g) shaved truffles.

3. The pureed mixture is poured and spread onto a lightly oiled pan to a thickness of $\frac{1}{8}$ inch. Refrigerate until set.

Using Color Sheets

The very thin, gelatin sheets are now ready for use. A set of small cutters, specially designed for minute decorations, is necessary. Some cutters are geometric shapes, others are letters, animals, leaf shapes, or other designs. Freehand decorations can also be very effective.

After pressing a selected cutter against an aspic sheet, the shape can be lifted with a toothpick and placed on cold fish, cold meat, or other cold food.

It is customary to cover the food to be decorated with one or two coatings of chaud froid, a cold cream sauce or velouté sauce with added gelatin to set the sauce firm when cold. For more details on the subject refer to *The Art of Garde Manger* by F. Sonnenschmidt and myself. A chaud froid has the advantage of preserving the freshness of the food to be decorated and also offers a neutral white background ideal for setting off color sheets.

Preparing the Designs

It is possible to prepare elaborate designs several days in advance. The first step consists of pouring the chaud froid onto 8- to 10-inch round plates or any other suitable dish. When the sauce has set, arrange the color sheet cutouts on the chaud froid. A decorative design on the edge of the circle can then be added. Seal the completed design with a thin coat of cold liquid gelatin. Refrigerate the plate, then lift the chaud froid from the plate and transfer it onto the food to be displayed, such as ham, turkey, galantine, or cold fish.

Cold ham is a favorite cold food display. It offers a large flat surface for decorating and is often seen on many elaborate buffets. Cold fish also offers many decorating possibilities. Colorful fish and shellfish, in various sizes and shapes, are easily adapted to decorative work. The pink color of salmon, the pure white of bass, tilefish, halibut, lake whitefish, cod, haddock and dover sole, and the coral red of several crustaceans provide numerous color contrasts when used with other ingredients. However, regardless of the motif or theme selected, simplicity is the rule for seafood decorations. These can be

1. The cucumber fish scales.
2. A truffle sailboat cut freehand and set on chaud froid, then transferred onto a poached, cold fish.
3. A lobster, cut out of orange sheet and set on a color contrasting fish or fillet.
4. Fish cut out with a fish cutter, using a red or orange color sheet.

87

Truffle sailboat.

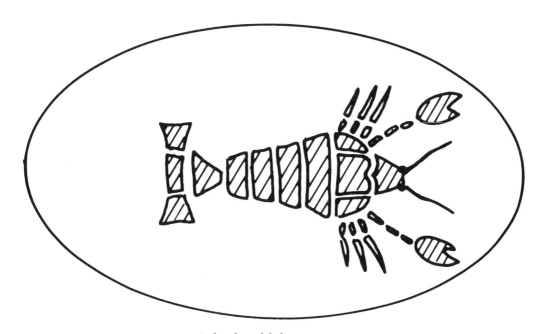

A freehand lobster cutout.

Many more color sheet designs can be created from the most simple to the most complete.

Also, the three-dimensional optical illusion is a favorite decorative design. It consists of three color sheets, orange, black, and yellow, arranged into a cube. Simply use a diamond aspic cutter and assemble the colors.

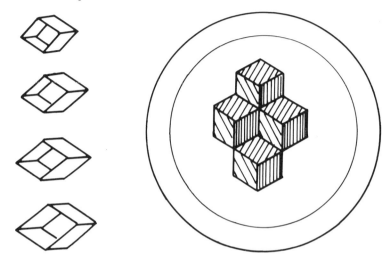

Another appealing method of decorating with color sheets is the angel fish in seaweed. Carve several angel fish freehand with a small paring knife. Select the most suitable color sheets such as the black, orange, red, and yellow. Arrange the cutouts directly on the food (mousse, whole cooked fish, fillet, or steak). For a seaweed effect, melt some green sheet and carefully spoon it around the angel fish.

5
Fanciful Meringues

A snowy mountain of egg whites shaped into swans, mushrooms, nests, baskets, and many other edible forms will delight any dessert lover. Meringue, simply made with egg whites and sugar, is an incomparable cholesterol-free modeling medium, delightfully sweet and easy to play with for anyone who likes to make fanciful creations.

Easy to make and to bake, meringue is prepared in a short time and dried at low temperature. The term *meringue* covers all forms of beaten egg whites and sugar, regardless of the proportions used.

The greatest care must be taken to keep all utensils and materials used free from grease. No trace of egg yolk should be allowed in the whites as grease will shorten the protein strands of the beaten egg whites and prevent air from being trapped within. A perfect meringue should be smooth, thick, fluffy and easy to handle.

Preparing Meringues

Making meringue is simple. For a standard recipe, the ingredients are

- 1 cup egg whites
- 2 cups granulated sugar

The quantities can be increased proportionally according to the amount desired.

1. In a mixing bowl, combine the egg whites with one cup of sugar. Whip the mixture using an electric beater until the meringue forms stiff peaks.

2. Gradually add the remaining sugar and any flavorings desired. Beat the meringue until it is smooth and thick.

Meringue is dried rather than baked. The ideal temperature should be 250 to 275°F (120 to 130°C). Higher temperatures will caramelize the sugar or alter the color of the meringue. The drying time in the oven will depend upon the size of the meringue sculptures. Small designs such as mushrooms and stalks will dry within an hour or two; whereas a large basket will need 4 hours to dry. Dried meringue does not spoil and can be stored in a dry place for several weeks without deteriorating in taste.

Tools Required for Meringue Decorations

A medium pastry bag and a set of decorative tubes are sufficient to create an array of designs with meringue. A pastry bag fitted with a large star tube can turn out richly puffed shells, rosettes, and other creations with rippled effects. Many other large tubes called drop flower tubes give excellent results by just squeezing a small amount of meringue onto a baking sheet covered with wax paper. Plan on having several pastry bags handy when piping various decorative effects with meringue or fit the bag with a coupler allowing the use of interchangeable tubes.

Simple round tubes can also give remarkable results. They are essential for mushroom meringues, the exotic meringue, and for shaping the neck of the meringue swan. The adaptable meringue lends itself to a multitude of decorations that are only limited by the imagination of the decorator.

 ## Meringue Fancies

Tools

    a medium size pastry bag
    a ½- to ¾-inch plain tube

1. Color the meringue with a touch of red food coloring. Mix well and it will turn pink.

2. Spoon the meringue into the pastry bag and squeeze out small round shells on a baking sheet lined with wax paper, following method described for mushroom meringue tops.

3. Dry at 250°F (120°C) for 1 to 1½ hours.

4. Spread the base of each meringue with a small amount of fruit preserve or melted sweet chocolate and make sandwiches.

These dainty desserts are delicious for snacks, with tea, or for a cocktail party.

The Exotic Meringue

Tools

> a medium size pastry bag
> a ½- to ¾-inch plain tube

The exotic meringue is an unusual and delectable dessert that is easy to prepare for any occasion.

1. Prepare a meringue recipe with banana flavor and a touch of yellow food coloring.

2. On a baking sheet covered with wax paper, pipe out the meringue mix into the shape of a banana using an up and down motion to obtain a rippled effect.

3. Dry in a 270°F (130°C) oven for 2 hours.

4. Dip the bases of the meringues in melted sweet chocolate. When set, spread a small amount of whipped cream over the chocolate.

5. Peel and slice a banana lengthwise. Trim it to the size of the meringue shells. Sandwich the banana slices between the prepared meringue bases.

 # Mushrooms on a Bed of Moss

Tools

> a medium size pastry bag
> a ½- to ¾-inch plain tube
> a fine strainer for the moss

1. Use a plain uncolored meringue. If desired add vanilla extract for flavor.

2. Spoon the meringue into the bag. Squeeze out plain round shells resembling mushroom caps on a baking sheet covered with wax paper. The size of the shells may be from ½ inch to 1 inch in diameter. To obtain perfect round tops, level any points with a wet brush. Sprinkle the mushroom tops with a small amount of cocoa to give a mottled effect.

3. To form the stalks, pipe out mushroom stems on the same baking sheet lifting the bag to form a peak.

4. Bake at 250°F (120°C) until dry.

5. To assemble, pierce a small hole in the center of the base of each top. Insert the pointed end of stalk into the meringue top to complete the mushroom.

6. The bed of moss is obtained by mixing almond paste with confectioner's sugar and a drop of green food coloring. Roll the almond paste into small balls and press through a fine wire strainer to get the effect of green moss. Scrape with a spatula and arrange around a cluster of mushrooms.

Decorate yule logs or Christmas cakes with a field of mushrooms on a bed of green moss.

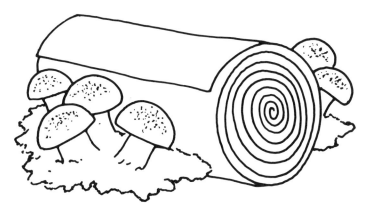

Meringue Swans

Tools

    2 medium size pastry bags
    1 medium flower tube
    a ¼-inch diameter plain round tube
    a ½-inch diameter star tube

A little practice and savoir-faire is necessary to make perfect swans. If at first the figures are not piped with satisfaction, scrape out the meringue and begin again. Squeeze out the meringue on a baking sheet covered with wax paper.

1. To shape the swan's head fit the bag with the round tube. Squeeze out a small dot for the head and, simultaneously, move tube away to shape a pointed bill. Touch the tube to the back of the head and in one continuous motion, squeeze to connect a curved neck, moving the tube in a question-mark-like motion.

2. To shape the body fit a pastry bag with a star tube. Squeeze out 3- to 4-inch oval shaped meringues. Fill the center to level. Squeeze out an edge around about 1 inch high.

3. To shape the left wing fit the pastry bag with the flower tube. Pipe out a 5-inch curved line with the right end pointed. With the narrow end of the tube pointed up, pipe out a feathery effect from right to left in an up and down motion. See the illustration for details.

4. To shape the right wing reverse the method.

5. Dry all the meringue shapes in a 250°F (120°C) oven for 1 hour.

6. To assemble spoon some ice cream or whipped cream into the oval shaped body of the swan. Stick the left and right wings on each side of body and secure the neck at the base of the wings.

Chocolate Meringue Designs

Tools

a large pastry bag
a medium to large star tube

1. Prepare a meringue recipe and fold in about 2 squares of grated sweet chocolate.

2. Fit the pastry bag with the star tube. On a baking sheet covered with wax paper, pipe the meringue into fingers, shells, half moons or any other design.

3. Bake at 250°F (120°C) for 1 to 2 hours.

Dip in melted chocolate if desired. Before baking, you may sprinkle some grated chocolate over the designs.

Coconut Meringue Nests

Tools

> a medium size pastry bag
> a medium size star tube

1. Fit the star tube into the pastry bag and pipe out a ring on a baking sheet covered with wax paper. Fill center of ring to level.

2. Build a spiral on the edge of ring to form a nest.

3. Sprinkle toasted coconut flakes all over the nest.

4. Dry in a 250°F (120°C) oven for two hours.

During Easter season, place Easter eggs in the nests or scoop ice cream into the nests.

 # Meringue Rocks

Tools

> a medium size pastry bag
> a plain tube

1. Color and flavor a meringue mixture. Fold in some toasted coconut flakes.

2. Cover a baking sheet with wax paper. Spoon the meringue into the shape of small rocks.

3. Dry at 260°F (125°C) for 1 hour.

Meringue Baskets

Tools

> a medium size pastry bag
> a medium size star tube

1. To form the base, follow the method described for coconut meringue nests. In a continuous motion, squeeze out a half ring for a handle.

2. To assemble, arrange some berries or ice cream in meringue base. Place handle on top to resemble a basket.

Larger baskets can be shaped in a similar fashion. Use a large star tube for best results. Build the edge of base 2 to 3 inches high in order to contain a larger amount of fruit or other ingredients. A large meringue basket filled with blueberries, raspberries, and strawberries becomes a glamorous dessert for any occasion.

 # Molded Meringues

Tools

> several small cookie molds
> a small spatula

1. Prepare a meringue recipe. Color and flavor if desired.

2. Brush the cookie molds with melted butter. Spoon in the meringue and spread level with a spatula.

3. Bake for 5 minutes at 270°F (130°C).

4. Unmold onto a baking sheet covered with wax paper and dry at 250°F (120°C) for 1 to 2 hours depending on the size of the molds.

5. Decorate the dried meringue with pieces of candied fruit or jams.

Meringue Sandwiches

Tools

> a medium size pastry bag
> a ½-inch plain round tube

1. On a baking sheet covered with wax paper, pipe out 4- to 5-inch-long meringue fingers attached four by four. Bake at 250°F (120°C) for 1 hour.

2. Sandwich 2 or 3 together with layers of ice cream or whipped cream.

3. To vary the presentation of the meringue sandwich, sprinkle with sliced almonds before baking.

Meringue Shells

Tools

> a medium size pastry bag
> a medium size star tube

1. Line a baking sheet with wet brown paper.

2. Pipe out 5- to 6-inch-long meringue shells using an up and down motion.

3. Dry in a 270°F (130°C) oven for 1 hour.

4. Cool and scoop out the bottoms of the shells. Fill them with fruit or ice cream.

More Simple Meringue Decorations

- ♦ Cover a cream pie with meringue stars and brown in a hot oven.
- ♦ An assortment of ice creams scooped on a serving platter and decorated with scrolls of meringue becomes an interesting dessert when browned under the broiler.
- ♦ Pipe out numbers and letters. Dry them at 250°F (120°C) and decorate birthday and anniversary cakes.

With meringue the possibilities are almost limitless. This is what makes the snow white mixture so appealing to food decorators.

Meringue mushrooms. *Background:* pastry bag. *Center:* ½-inch plain tube. *From left to right:* meringue mushroom caps and stems; mushroom cap pierced to hold the stem; the assembled meringue mushrooms.

Notice the versatility of the star tube. It makes shells, fingers, half moons, rosettes.

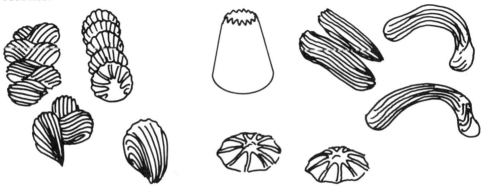

FANCIFUL MERINGUES 99

Meringue creations made with a plain pastry tube. *Left from top down:* letters of the alphabet for cakes. *Top from left to right:* meringue sandwich; meringue fancies; meringue rocks; meringue numbers for cakes. *Center left:* rippled meringue sandwiches. *Bottom right:* the exotic meringue.

Meringue is an extremely versatile dessert medium. With a little imagination you can create new designs or adapt the decorations presented in this book. Here are some interesting possibilities.

- ♦ Make nests of coconut meringue and fill them with different colored grapes and berries.
- ♦ Dip meringue rosettes, fingers, shells, and rocks into chocolate.
- ♦ Add a handle to a meringue nest to form a basket. Then fill the basket with berries. (When making a basket, pipe the handle separately and attach it when dry.)

🌿 6 🌿
More Food Decorations

Butter and Cream Cheese Creations

Butter and cream cheese used separately or mixed are mostly used when the application must be done with the help of a pastry bag. The shape and patterns of the piped product will be determined by the type of metal tube set in the pastry bag.

Cream Cheese Roses

Tools

    nails mounted on 1-inch-diameter discs or ¼-inch dowels 5 to 6 inches long
    a pastry bag
    a rose pastry tube

1. Fit the pastry bag with the rose tube. Spoon the softened cream cheese into the bag. Pipe the cream cheese in sequences as shown. Each petal should be piped starting at the base of the flower and following a half circular motion. If several roses are required, it is preferable to pipe the rose bud using a plain pastry tube, then build the rose petals around the bud as explained.

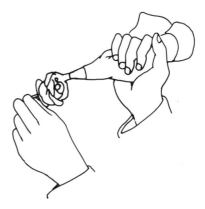

2. Another simple and very effective way to create cream cheese roses is to build the rose around the tip of a dowel. Lift the flower from the dowel and refrigerate on wax paper.

3. The cream cheese may be colored with a touch of yellow or red food coloring.

Use the cream cheese roses as appetizers on crackers. They can also be displayed with smoked salmon, meat platters, and many other food displays.

Butter Decorations

Butter molds of all sizes and shapes are available on the market. Wooden molds are more dependable than those made of plaster or other materials. Individual servings of butter or a large display of butter decorations for a crowd are always welcome decorative details to a simple meal or an elaborate buffet.

To obtain the best results with butter decorations, follow the directions that accompany the molds. The illustration in the color insert shows butter decorations in the form of a bird, a fish, and a fruit mold.

Bread Decorations

Many of the most useful decorative centerpieces consist of fancy loaves of bread. Any size or shape loaf can be baked. Pumpernickel, crusty golden french loaves, whole wheat or rye sandwich loaves, loaves studded with sesame or poppy seeds, and many others add a pleasant color contrast especially when displayed with other foods. The use of bread in food decorating can go a step further. The following methods are innovative and unusual.

Bread Crepe Pans

Tools

a cast iron crepe pan
pie crust dough
a paring knife

1. Make a firm pie crust dough without fat. Roll the dough to ¼-inch thickness. Cover the cast iron crepe pan including the handle. Press the dough against the bottom and the handle of the pan. Cut out any excess of dough overlapping the crepe pan.

2. Bake the dough in the crepe pan at 350°F until golden brown. Part of the handle may be cooked before the remaining crust, in which case, it should be covered with foil to prevent additional browning.

3. Cool the cooked bread pan then take it out of the crepe pan.

Use the cooked bread pan to serve cold vegetable salads, scrambled eggs, sausages, and meat pies. The handle is what makes this decorative bread display so unusual.

Bread Basket and Horn of Plenty

A bread basket filled with fruit or a cornucopia displayed on a buffet can be beautiful when well done. Instead of rolling the stiff bread dough by hand, it is put through a sausage machine. The dough comes out uniformly thick and smooth. In fact, it is so easy to use that a large basket can be formed in a short time. Rolling the dough by hand is time consuming and seldom gives great results.

A Cornucopia

Tools and materials

chicken wire
aluminum foil
a sausage machine attachment
a non sticky bread dough

1. Cut and shape the wire into the form of a horn of plenty. Cover the surface with heavy foil. Place on a baking sheet.

2. Secure the sausage machine attachment to the electric mixer. Put the bread dough through the sausage attachment. The mixer should run at low speed.

3. Cover the whole surface of the horn of plenty with coils of dough. Brush the dough with beaten eggs and bake at 375°F until the dough is golden brown and cooked.

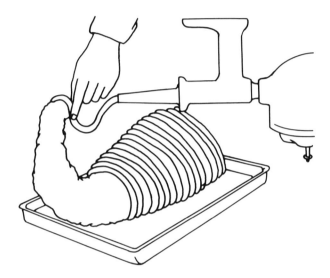

Filled with fruit and nuts, the horn of plenty is ideal for a Thanksgiving display.

Bread Baskets

Tools and materials

a perforated aluminum board, one side fitted with round perforations, the opposite side fitted with oval perforations
¼-inch dowels.
a non-sticky bread dough

1. Cut the dowels into 6- to 8-inch-long sections. If a handle is needed, soak a long dowel in hot water so that it becomes pliable and can be fitted on the perforated board as shown. If the basket has a handle, make sure it fits in the oven! Select the desired size of the basket and arrange the dowels into the perforations as shown.

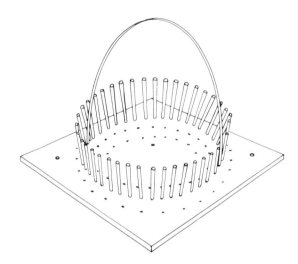

2. Put the bread dough through the sausage machine attachment with the mixer at low speed.

3. Weave the dough around the dowels. Build up the basket to the height of the dowels so that they are concealed.

4. Brush the basket with beaten eggs and bake at 375°F until golden brown and cooked.

MORE FOOD DECORATIONS 107

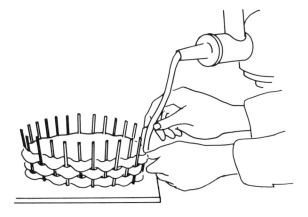

5. Lift the bread basket off the board leaving the dowels intact. If a bottom is desired, it can be made on a separate baking sheet and then fitted inside the basket after it is baked.

Fill the basket with loaves of bread, rolls, or fruits. Display it as a centerpiece on a buffet or a dining table.

Eggs in Food Decorating

Eggs have always exerted a great fascination to humankind and stimulated the imagination of cooks and chefs. Eggs should be part of a balanced diet. They contain proteins, fats, and minerals. The importance of eggs in our kitchens cannot be overemphasized. Always buy the freshest and the best quality eggs. Many dishes can be created using eggs. From a decorative standpoint, hard boiled eggs can be transformed into a myriad of intricate garnishes.

 ## The 100-Year-Old Eggs

Boil eggs for 5 to 6 minutes. Cool the eggs, crush each shell lightly but do not remove any pieces. Immerse the eggs in a strong tea solution kept at boiling point and flavored with ginger or anise. Marinate eggs for 1 to 2 hours, chill, remove shells, and serve. The marble streaks that penetrate into the cracked eggs give an antique look to the eggs. The taste is not altered. Slice off the bases so eggs stand upright.

 Egg Mimosa

The mimosa garnish consists of hard cooked egg whites and yolks chopped separately. Chopped parsley is added. The combination of green, white, and yellow sprinkled over all sorts of salads is a simple colorful garnish.

Egg Slices

An egg slicer is a useful gadget in any kitchen for slicing hard cooked eggs. Slicing hard cooked eggs with a knife is a difficult task. Sliced cooked eggs can be used as a garnish on many types of salads. They can be displayed in alternating rows with tomato slices or in a ring.

The Chinese Farmer

This is an eye catching presentation for hard boiled eggs.

1. Cut a sliver off the base from the thicker end.

2. Top with a round, cooked potato or white turnip ball cut with a Parisian scoop. Secure with a toothpick.

3. The hat is a cone of salami sliced thinly.

4. The eyes and buttons are cutouts of either black olives or truffle sheet dipped into liquid gelatin to stick to the eggs.

5. Two toothpicks extend on each side with a mushroom pail dangling at each end.

The Snowman

1. Cut a slice off the thicker end of the egg to form a base.

2. The head can be either a round, cooked potato ball or a turnip ball cut out with a Parisian scoop. Secure with a toothpick.

3. The hat is a slice of egg white. It can be the slice cut off the base and secured on the head with a toothpick.

4. A scarf surrounds the neck of the snowman. It is simply made of a strip of blanched green leek.

5. The eyes and coat buttons are truffles or ripe olive cutouts dipped in liquid gelatin to better stick on the egg.

Stuffed Whole Eggs

1. Cut a slice off the base so that the egg will sit on its side.

2. With a large Parisian scoop, carefully remove a round piece of the top of the egg. Fill the cavity with either red or black caviar, or a cherry tomato. Decorate with a truffle sheet.

The Lady Bug

1. Cut a hard boiled egg in half.

2. Cover $\frac{3}{4}$ of the surface of the egg with a round piece of pimiento sheet.

3. Use black dots of truffle sheet cutouts to decorate the top. The legs are also truffle cutouts.

More Decorative Ideas

 ## The Liver Pâté Cone

1. Shape about a pound of liver pâté into a ball tapering down at the base.

2. Use either whole, blanched or unblanched almonds, or slivered almonds and stick them into the liver pâté.

3. Cover the entire surface of the pâté with the almonds. Chill the cone and serve as an appetizer with crackers.

 ## Snow Peas Stuffed with Cream Cheese

Tools

a small paring knife
a pastry bag fitted with a star tube

MORE FOOD DECORATIONS   111

1. Select fresh crisp snow peas. Soak them in cold water if additional crispness is required. Cut off the tips of the snow peas.

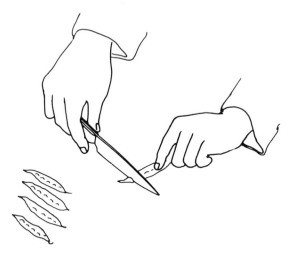

2. Open one end of the snow peas to form a pocket.

3. Pipe the soft cream cheese into the pocket. Sprinkle with dill for color.

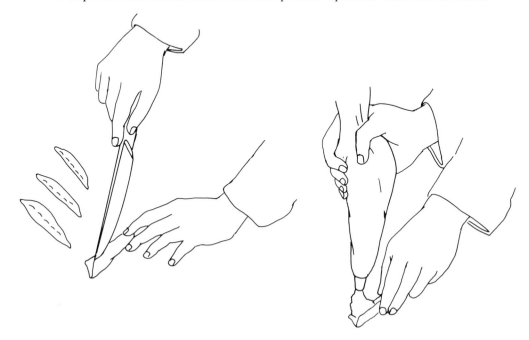

Stuffed snow peas make an excellent appetizer.

Thanksgiving Turkey Appetizer Tray

Draw an outline of a turkey on $\frac{1}{4}$-inch plywood board. Carve the board with an appropriate saw. The appetizers should be on a bread base and are cut to match the shape of the turkey. Canapes made of red and black caviar, egg yolks, and egg whites make a colorful Thanksgiving appetizer tray.

7
Bouquets and Containers

Bouquets of flowers and food containers add interest to a platter or a buffet table. A vegetable bouquet can be as simple as radish flowers held on toothpicks and presented in onion shells, or secured to a head of lettuce or a potato base and garnished with parsley sprigs. Carrot flowers, such as spider mums, carrot lilies, and other vegetable flowers can be combined and displayed on a cabbage or a lettuce base, held by toothpicks, and served as appetizers.

Some containers are perfect sauce boats, such as the cucumber boat; the apple, orange, or grapefruit tulip; and the green pepper basket. Others are used to serve individual portions of foods, like the avocado shells, papaya halves, and lemon or orange shells.

Baskets can be made of green, red, or yellow peppers; apples; lemons; oranges; and other fruits. These decorative containers can be used to serve all sorts of food.

Pineapple shells, a bread basket, bread pans, and potato nests can be filled with many different foods and displayed on food platters or on buffets. A potato nest garnished with quail eggs and roasted or braised quails is a visual delight.

Last but not least, a bouquet arrangement worth displaying is shown in the following section. The butternut squash vase filled with daisies represents the ultimate in food decorating. Other flower arrangements can be created in the same vase. The ornateness of the bouquet does not depend on the budget, but on the imagination of the creator. There should not be any waste. Leek, radish, carrot, and other flowers can be combined in the butternut squash vase and used as a centerpiece for cocktail parties, buffets, or other festive occasions. Whatever the method used to present any flower decorations, make it simple, uncrowded, and appealing to the eye.

 A Bouquet of Daisies

Tools and materials

a butternut squash
3 or 4 white turnips, peeled and $\frac{1}{2}$ of a yellow turnip
one carrot, peeled
one leek with green tops

a bunch of scallions
wire stems
a large paring knife
a lemon peeler
a small Parisian scoop
2 daisy cutters of different size

1. Assemble all the materials and tools listed.

2. Slice off the top of the squash.

3. Scoop out the inside of the squash.

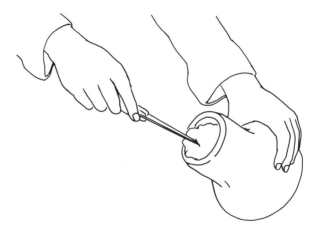

4. Decorate the top portion of the squash using the lemon peeler.

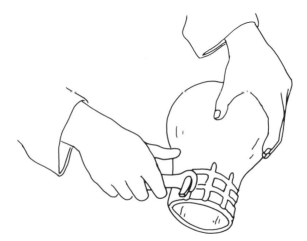

5. Using a small Parisian scoop, alternate a row of carrot balls and turnip balls around the circumference of the squash.

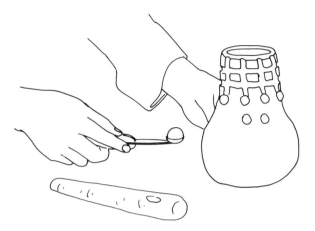

6. Insert the pieces of wire into the scallion stems. Arrange the stems inside the squash.

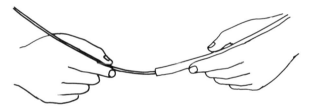

7. Slice the turnips into ⅛-inch-thick slices. Cut out the flowers with the decorative cutters.

8. Assemble the flowers, joining two flower cutouts. Stick flowers onto the stems. Decorate the centers with balls of squash. Arrange the flowers in the butternut squash vase.

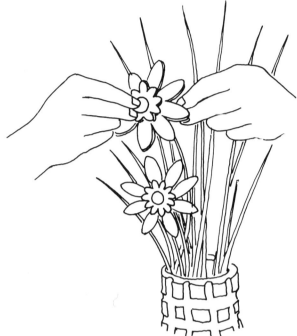

9. Trim the green part of the leeks for foliage and arrange them in the vase.

A combination of radish flowers with turnip daisies make a colorful flower arrangement. Display the flowers in their vase on a buffet or on a dining table as a centerpiece.